Unravelling desertification

To **John B. Thornes**
*whose vision, enthusiasm and excellence in research
has changed the way we view, conceptualise
and analyse desertification problems*

Unravelling desertification

Policies and actor networks in Southern Europe

Edited by
Geoff A. Wilson
Meri Juntti

Wageningen Academic
P u b l i s h e r s

Subject headings:
deseertification
policy implementation
actor networks

Photo cover by Neil Roberts

ISBN 9076998426

First published, 2005

Wageningen Academic Publishers
The Netherlands, 2005

Table of Contents

List of figures

List of tables

List of contributors

Nikos Beopoulos is Professor in the Department of Rural Economics and Development at the Agricultural University of Athens (Greece). His research focuses on EU agri-environmental policies and the relationship between agriculture and the environment. He has authored various books and articles in international journals on these research topics.

Peter Eden is a researcher at the Institute for Rural and Environmental Management (Instituto para o Desenvolvimento Rural e Gestão Ambiental) in Lisbon (Portugal). His research has focused on policy and conservation issues in the Portuguese countryside, and he has published several articles and book chapters on these research topics in both English and Portuguese.

Daniele Ferraretto obtained his first Masters degree in Forest Sciences at the University of Padova in 2000, and he has just completed his second Masters degree (MA Environment, Politics and Globalisation) at King's College London, UK. He is currently a Research Fellow at the Italian National Institute of Agricultural Economics (INEA), and also participates in the research activities of the Desertification Group at INEA. His research activities focus on the analysis of social and economic aspects affecting desertification processes in Southern Italy, and social and economic characteristics of irrigated agriculture in north-eastern Italy.

Meri Juntti, the co-editor of this book, was a postdoctoral research fellow in the Department of Geography, King's College London, UK, and is currently a senior research associate at the Centre for Social and Economic Research on the Global Environment (CSERGE) at the University of East Anglia (UK), where she is working in the research programme on 'Environmental Decision Making'. Her research focuses on policy mechanisms guiding European environmental conservation, agriculture and rural development from an actor-oriented approach, with empirical interests centering on European agri-environmental policy and on environmental integration into a wider set of land management related policies at the EU level.

Juan Oñate is a full-time lecturer in Ecology at the Department of Ecology in the Universidad Autónoma de Madrid (Spain). Besides having participated in research projects funded by the Spanish administration and the European Commission, he has also wide experience in applied projects related with environmental planning and environmental impact assessment. His research interests are currently related to the linkages between agricultural policies and nature conservation, as well as strategic environmental assessment. He has authored or co-authored more than 30 research papers and book chapters.

Begoña Peco is head of the Terrestrial Ecology Group at the Department of Ecology in the Universidad Autónoma de Madrid (Spain). She currently acts as the Land Use Cover Change Focal Point of the IGBP-Spain. Her research in community ecology has particularly covered grassland species richness and vegetation structure in relation to the physical environment and anthropogenic perturbations. At the landscape level, her interest focuses on temporal and spatial variations in landuse and their relationship to different physical and socio-economic variables and the influence of agricultural policy. She has authored or co-authored more than 80 research papers and book chapters.

Andrea Povellato is a senior researcher at the Italian National Institute of Agricultural Economics (INEA, Italy) since 1990 and was an Associate Lecturer in Agricultural Economics at the University of Padova (Italy). His research activities have focused on issues concerning agriculture-environment relationships, with studies on rural development, nature protection, environmental impact assessment, innovation diffusion and agricultural extension services. During the last few years he has been involved in several research projects concerning the evaluation of agri-environmental policies and the use of agri-environmental indicators (OCSE workshops). He is also an active member of the Desertification Group at INEA and has published widely in various international journals and books.

Miguel Vieira is a senior researcher at the Institute for Rural and Environmental Management (Instituto para o Desenvolvimento Rural e Gestão Ambiental) in Lisbon (Portugal). His research has focused on policy and conservation issues in the Portuguese countryside, Portuguese agri-environmental policy, and implications of Portugal's integration into the EU agricultural policy-making framework, and he has published widely on these research topics in both English and Portuguese.

George Vlahos is an agronomist with an MSc in Rural Development, and is currently employed as a researcher in the Department of Rural Economics and Development at the Agricultural University of Athens (Greece). His main research activities revolve around EU agri-environmental schemes, agricultural and rural development policies, as well as investigating environmentally-friendly farming practices.

Geoff Wilson, one of the editors of this book, is Professor in Human Geography at the School of Geography, University of Plymouth (UK). He works mainly on issues related to European environmental policy and politics and agriculture and conservation, and has published widely in the field. He has wide experience with regard to international research projects and book editing. Main books include, for example, a text entitled 'Environmental management: new directions for the 21st century' (with Raymond Bryant) published in 1997 by UCL Press, an edited book on European Union agri-environmental policy (with Henry Buller and Andreas Höll) published in 2000 by Ashgate, and an analysis of 'German agriculture in transition: society, policies and environment in a changing Europe' (with Olivia Wilson) published in 2001 by Palgrave. In addition to this, Geoff Wilson, has published over 40 articles in key international geographical and environmental journals.

Acknowledgements

This book could not have been successfully produced and completed without the help of many people and institutions. Authors in this book would like to thank the European Commission for the funding of Module 2 of the MEDACTION project upon which this book is based. Particular thanks must go to John Thornes (to whom this book is also dedicated) for initiating the MEDACTION project (which, together with the DESERTLINKS project, formed a continuation of the long-running EU-funded MEDALUS projects), and for bringing in social scientists to complement already existing natural sciences work on desertification in Southern Europe. We also wish to thank Jane Brandt and Nicola Geeson for the immense amount of work involved in making both the MEDACTION and DESERTLINKS projects a success, and for their enthusiasm and dedication with regard to furthering desertification research in Southern Europe.

The editors also wish to thank the four research teams from Spain, Portugal, Italy and Greece for their excellent cooperation in the collection of empirical material at case study level in the four case study areas used in this project, and for their contributions to MEDACTION reports and to this book. Many thanks must also go to Dolf de Groot (University of Wageningen, The Netherlands) who did an excellent job in coordinating the MEDACTION project, ensuring that deadlines were met, and overseeing the production of reports to the Commission. Dolf, particular thanks for your patience and your constructive criticism of our work throughout the duration of the MEDACTION project. Thanks must also go to Anton Imeson and Mike Kirkby for challenging discussions surrounding the issue of desertification - discussions that have also partly found reflection in some of the arguments made in this book. We also would like to thank Tim Absalom, University of Plymouth, for help with drawing some of the figures used in this book.

Last, but certainly not least, we also wish to thank Mike Jacobs and colleagues from Wageningen Academic Publishers for accepting to publish this book and for comments and help during the final stages of book production.

List of abbreviations

AEP   Agri-environmental policy
ANT   Actor-network theory
AWU   Annual working units
BSE   Bovine Spongiform Encephalopathy
CAP   Common Agricultural Policy
CASMEZ   Fund for the Development of Southern Italy
CHS   Confederación Hidrográfica del Segura
CMO   Common Market Organisation
COP   Conference of the Parties (UNCCD)
CRIC1   Committee for the Review of the Implementation of the [UNCCD] Convention (first
         meeting)
CSF   Community Support Framework (Portugal)
CST   Committee on Science and Technology (UNCCD)
EC   European Community
EEC   European Economic Community
ESAs   Environmentally sensitive areas (desertification-related)
EU   European Union
GAEC   'Good agricultural and environmental conditions' (CAP)
GEF   Global Environmental Facility
ICCD   Italian Committee to Combat Desertification
INCD   International Convention to Combat Desertification (UNCED)
LFAs   Less-favoured areas (CAP)
LRSs   Land Reclamation Syndicates (Italy)
LUs   Livestock Units
NAPs   National Action Plans (UNCCD)
NCUs   National Coordinating Units (UNCCD)
NFPs   National Focal Points (UNCCD)
NGOs   Non-governmental organisations
NHP   National Hydrological Plan (Spain)
NUTS   Nomenclature of territorial units for statistics (EU)
OECD   Organisation for Economic Cooperation and Development
PACD   Plan of Action to Combat Desertification (UNCOD)
PAMAF   Programme of Support for Modernisation in Agriculture and Forestry (Portugal)
PFO   Professional Farmers' Organisation (Spain)
RAPs   Regional Action Programmes (UNCCD)
RDR   Rural Development Regulation (CAP)
UAA   Utilised agricultural area
UN   United Nations
UNCCD   United Nations Convention to Combat Desertification
UNCED   United Nations Conference on Environment and Development (1992)
UNCOD   United Nations Conference on Desertification (1977)
UNEP   United Nations Environmental Programme
WCED   World Commission on Environment and Development

Part I

Setting the scene: desertification and policies in Southern Europe

This book will be comprised of three distinctive parts that aim at providing background information, empirical case study evidence, and a theoretical investigation of actor networks operating in the context of desertification policies in Southern Europe.

The first part of this book sets the scene for understanding the complex interlinkages between desertification and policies and is comprised of two chapters. The introduction (**Chapter 1**) introduces the reader to desertification issues and definitions and makes a case for increased inclusion of the 'human dimension' to desertification research using an actor-oriented approach. **Chapter 2** discusses in-depth issues related to desertification and policies ranging from the global to the European to the national arena. Particular emphasis is placed on the *United Nations Convention to Combat Desertification* (UNCCD) policy framework and the problems of transposing global policy frameworks to local desertification problems, as well as on highlighting that many policies can act both as drivers and mitigators of desertification.

1. Introduction

M. Juntti and G.A. Wilson

1.1 Setting the scene

Although humans have used the Earth's land surface for cultivation for thousands of years, the issue of possible negative effects on the regenerative capacity of soils has only recently become a globally acknowledged and debated issue (Blaikie and Brookfield, 1987). Often, an irreversible decline in the soil's reproductive capacities has been termed 'desertification' - a term that refers to accumulative detrimental processes consisting of several factors, usually categorised into human and natural causes. While the term 'desertification' was first introduced into the scientific literature as early as the mid 20th century, global scale recognition of the problem only emerged after the United Nations 1977 *Desertification Conference* in Nairobi (Middleton, 1995). However, resulting actions and policies to alleviate desertification processes have only been implemented very slowly (e.g. European Commission, 2000), and interpretations of what exactly constitutes the 'problem of desertification' continue to remain vague and varied, in particular among sub-national and grassroots actors and institutions (Rubio, 1995).

The term 'desertification' in itself is somewhat misleading. Whereas, as a biological concept, 'desert' implies a characteristically dry, sterile and unproductive area, a desert can also refer to wilderness - an often highly valued area characterised by lack of inhabitation and isolation (Rubio, 1995). Similarly, if 'desertification' is understood as a cumulative effect of land degradation through erosion, it has proven very difficult to define through empirical scientific exploration whether an area is suffering from desertification or not (Thornes, 1998a). This has partly been linked to the fact that field measurements have tended to be technique-dependent, and the extent of crop yield and general soil quality - key indicators in the assessment of the extent of desertification in a given area - are often difficult to measure due to lack of technology, equipment and long-term data (Middleton, 1995; Lal, 2001). As a result, in many areas the nature, extent, causes and effects of desertification are very controversial and hotly debated.

Often, 'desertification' is seen to result from over-intensive landuse methods, but rarely is it acknowledged that these human causes of desertification are embedded in economic, socio-cultural and political dimensions (Hannigan, 1995). As with most environmental problems which are socially constructed on the basis of varying interpretations of scientific knowledge (Demeritt, 1996), the consequent uncertainties surrounding the 'actual' extent of damage and need for action can easily become a political weapon (Blaikie and Brookfield, 1987; Castree and Braun, 1998). In particular, the continuing ambiguity and the resulting competition between different interpretations of 'desertification', and its possible ecological and economic impacts, have hindered large-scale and concise efforts to tackle the problem in practice (Glenn *et al.*, 1998).

The aim of this book is to address these issues and to 'unravel' the policy-related drivers of desertification. Using a 'human geography' approach, we address desertification both as a spatial concept surrounded by a multitude of different discourses, and as a tangible unsustainable process that is connected to a complex set of policies and changing land management practices. The focus of this book will be on Southern Europe, where desertification has been a long-standing problem in many areas, and where, in some places, the loss of productive capacity has worsened considerably over the last few decades (Thornes, 1998a, 1998b). By focusing on four specific case study areas in Portugal, Spain, Italy and Greece, the book will cover the 'human dimension' of desertification, exploring in particular how the framework of existing policies has affected land management decisions and desertification processes. The emphasis will be on how policies may have contributed to desertification alleviation and mitigation, as well as to a worsening of desertification processes. By using an *actor-network approach*, we will specifically investigate the importance of networks of actors that shape the nature and direction of policies that affect desertification processes. In this sense, this book aims at providing a first glance into the complex policy, economic and socio-cultural networks that operate at the local, regional and national levels in areas of Southern Europe affected by desertification, and to analyse how these networks hinder, or promote, the implementation of policies aimed at alleviating the threat of desertification.

1.2 Understanding 'desertification': definitions and debates

There are many different definitions, interpretations and conceptualisations of the term 'desertification', and heated debates have ensued between the different schools of thought. The *United Nations Convention to Combat Desertification* (UNCCD), ratified in Paris in June 1994, is by far the most significant existing policy framework addressing desertification at a global scale, and therefore provides a good starting point for understanding the problems of defining 'desertification'. The Convention has provided an overarching global policy framework that affects landuse decisions and national policies both in countries with only 'moderate' desertification problems (e.g. Southern Europe), as well as in regions of the world experiencing very serious drought and/or desertification conditions (e.g. large parts of Africa) (Middleton, 1995; Botkin and Keller, 1998). On a global scale, the Convention has not only enhanced awareness of the problem of desertification, but has also triggered international-scale projects and national-level action programmes to combat the problem in practice (see also Chapter 2). The European Union[1] (EU), for example, along with over 150 other international organisations and governments, has ratified the Convention which entered into force in 1996 (European Commission, 2000).

However, despite its undeniable focus on improved environmental management of already desertified areas and areas prone to desertification, the Convention has already had to face severe criticism. First, the definition of desertification used in the Convention as "land

---

[1] For reasons of consistency and simplicity we will use the term EU (European Union) throughout this book for any reference to the community of European countries, but we acknowledge that for any developments before 1987 the correct term should be 'EEC' (European Economic Community) and between 1987 and 1992 'EC' (European Community).

degradation in arid, semi-arid and dry sub-humid areas resulting from various factors, including climatic variations and human activities" (UN, 1994, 4), has been criticised as being too vague and ambiguous. Second, although human activities are mentioned in the definition, the emphasis is more on biological processes and technical aspects, thereby sidelining political, economic and socio-cultural dimensions of desertification. Similar problems are apparent in the UN's definition of 'land degradation', which does not sufficiently contextualise human activities in their broader political and socio-economic context, where 'land degradation' is seen as

"the reduction or loss, in arid, semi-arid and dry sub-humid areas, of the biological or economic productivity and complexity of rainfed cropland, irrigated cropland, or range, pasture, forest and woodlands resulting from land uses or from a process or combination of processes, including processes arising from human activities and habitation patterns, such as: (i) soil erosion caused by wind and/or water; (ii) deterioration of the physical, chemical and biological or economic properties of soil; and (iii) long-term loss of natural vegetation" (UN, 1994, 5).

Actions to combat desertification implied in the Convention are left equally ambiguous, and hence remain open to different political interpretations. Actions are described as "activities which are part of integrated development of land in arid, semi-arid and dry sub-humid areas for sustainable development which are aimed at: (i) prevention and/or reduction of land degradation; (ii) rehabilitation of partly degraded land; (iii) reclamation of desertified land" (UN, 1994, 4).

Other conceptualisations of 'desertification' have arguably used a more holistic approach that places more weight on political, economic and socio-cultural factors. One example is the *World Commission on Environment and Development* (WCED), established in 1983, which put more weight on the political dimension of desertification than the above instrumentalist definitions. There, desertification is seen as "the process whereby productive arid and semi-arid land is rendered *economically unproductive*" (WCED, 1987, 34; emphasis added). Yet, even in this definition the emphasis is on economic aspects of desertification at the possible expense of other socio-cultural factors. Nonetheless, this brief definition incorporates the socio-economic context of the locality in which desertification occurs, embedding the concept at the very least in the economic interactions in which the price of land and its products are negotiated. In this respect, the WCED definition is more or less consistent with the anthropocentric approach used by the *United Nations Environmental Programme* (UNEP), where the emphasis is on effects of desertification *on people* rather than on land (Glenn *et al.*, 1998).

While it is clear that the difference in emphasis detectable in the above definitions can lead to very different ways of conceptualising and diagnosing the problem of 'desertification', and, consequently, to the adoption of different remedial methods, different emphasis can also be used to serve different interests. As with all environmental issues, the desertification issue is not free of interest conflicts and ulterior motives. It can, therefore, be argued that successful implementation of environmental policy instruments like the UNCCD requires the exploration of the wider socio-economic context - both institutional and attitudinal aspects of the desertification issue. In this context, recent analyses of European agri-environmental policies (AEPs) have shown that institutional and interest-related aspects of both policy delivery and adoption behaviour can provide important explanations for policy failure or inefficiency, and uncover the needs for new knowledge and learning among both policy implementers and recipients (e.g. Morris and Potter, 1995; Wilson and Hart, 2001; Wilson and Buller, 2001; Juntti and Potter, 2002). Indeed, policies are interpreted differently depending on what kind of moral criteria and justifications are found for their execution (Lowe *et al.*, 1997). Thus, implementation gaps, resulting from evolving interpretations of policy in the implementation process and from the translation of policy into tangible practices in varying economic and social contexts, can often be understood as the strategies individual actors select in the pursuit of their own goals, whilst tackling the specific policy issue at hand (Sabatier, 1986; Hill and Hupe 2002). Yet, even if we acknowledge the importance of addressing a global issue like desertification at the actor level - in other words, the *human dimension of desertification* - we are faced with substantial methodological challenges. It is evident that a robust justification for such an anthropocentric approach is required by those who continue to advocate and perpetuate instrumentalist and positivist interpretations of the desertification problem (Long and Long, 1992) - an issue we address in the next section.

1.3 Addressing the human dimension of desertification: the EU-funded MEDACTION project

It is evident that in both research and policy, the relative roles of human and natural factors in causing the problem of desertification are subject to different interpretations (Redclift, 1987; Mannion and Bowlby, 1992; Middleton, 1995; Wilson and Bryant, 1997). For example, a common denominator among different definitions of desertification is that it is seen to take place in typically dry areas with high variability in precipitation, where the impact of human management practices largely depends on these variable natural conditions. Yet, these two processes can often be difficult to distinguish from each other (Atchia and Tropp, 1995). While the four main anthropogenic causes of desertification have generally been agreed as overcultivation, overgrazing, deforestation and soil salinization due to irrigation (Kinlund, 1996), it appears that explanations of cause and effect have failed the challenge of overcoming the dichotomy of natural versus social causes and consequences.

While this dichotomy raises interesting theoretical and methodological questions - some of which will be addressed in this book - it is not associated with academic enquiry alone. It is human detachment from nature that some environmental writers claim to lie behind environmental degradation, problems that can be interpreted as symptoms of imposing the rules of an accumulative socio-economic regime upon nature (Wilson and Bryant, 1997). It is in this context that approaching environmental issues from an actor perspective allows us to understand how the perceptions that we hold concerning the relationship between nature and society - whether seeing nature as serving economic accumulation or environmental sustainability needs - can perpetuate a detachment that culminates in the environmental problems we are currently facing (Eder, 1996; Castree and Braun, 1998). Yet, in the case of environmental degradation in general, and desertification in particular, the understanding of this moralistic interpretative aspect has not been aided by the traditional emphasis on natural science discourses, methods and technical solutions that are used by many researchers focusing on desertification issues. This, in turn, has led to a relative neglect by most of the literature investigating desertification processes of *thickly textured research* focused on policy formulation and implementation, landholder decision-making, and attitudinal and behavioural factors.

Europe, where desertification has developed into a problem mainly in the Mediterranean regions, serves as an example of this natural science interpretation of desertification. The *Fourth Annex* to the UNCCD identifies the European countries bordering on the Mediterranean Sea (Spain, France, Portugal, Italy and Greece) as countries with a marked problem of desertification, requiring the design and implementation of *National Action Programmes* to mitigate the negative effects of desertification (Thornes, 1998a). As a result, desertification has been featured in research frameworks of the European Union since 1989, the initial aim having been to achieve a better understanding of the problems and processes of the phenomenon in relation to natural hazards. Key research projects have included the *European Field Experiment in Desertification Threatened Areas Project* (EFEDA), focusing on climatological factors (Colomer *et al.*, 1995); the ARCHAEOMEDES project exploring the historical aspects of landuse change and their impact on the processes behind desertification (European Commission, 2000); the DEMON project developing an integrated approach to Mediterranean land degradation mapping and monitoring by remote sensing (Valor and Castellas, 1995); and the MEDALUS projects (I-III) which, during the 1990s, have analysed land surface processes that contribute towards desertification (Thornes and Mairota, 1995; Brandt and Thornes, 1996; Geeson *et al.*, 2002).

Yet, it has been increasingly recognised that the implementation of the UNCCD policy framework is facing very different problems from those studied in these projects. For example, while the UNCCD provides specific guidelines for the implementation of the Convention in desertification-affected areas in Southern Europe (UN, 1994, Annex IV), it appears that the initial (and often misguided) interpretation of the Convention within Southern Europe was that it referred to the development of cooperation between Southern Europe and Africa, rather than forcing Southern European countries to tackle desertification at their own doorstep (Mourao, 1998). Thus, while the *National Action Plans* (NAPs) are finally beginning to be drawn up, there are several relatively under-researched policy-related and economic struggles that influence the process of their implementation, and that will encourage different interpretations of the effectiveness of these programmes.

Thus, although the 'human dimensions' influencing desertification have started to receive more attention (e.g. Thornes, 1998b), the prevalent research and policy agenda in Southern Europe can be described as *reductionist* and *mechanistic*, manufacturing ready-made technical solutions designed to be implemented through a linear process of information and technology transfer. As a result, the main focus has been on the natural science processes behind desertification and on developing technological solutions to the ensuing problems. Indeed, natural scientists have tended to hold a monopoly on the research field, at the expense of social science enquiry into the role of human action (Middleton, 1995; see also Redclift *et al.*, 2000). Consequently, the socio-economic causes and consequences of desertification, and aspects that might influence possible mitigation measures, have largely been left without attention. We argue that this is a telltale sign of a top-down and structurally determinist approach in the design and implementation of policies to combat desertification (e.g. Marsden *et al.*, 1993; Murdoch, 1997).

There are, nonetheless, signs of a more inclusive approach, one of these being European funding granted to the MEDACTION project (January 2001 - March 2004), which has provided the empirical basis for this book. The MEDACTION project focused on policies for landuse to combat desertification, and analysed various aspects of the policy environment and the socio-economic context in which desertification is taking place in the four Southern European target countries of Portugal, Spain, Italy and Greece. While the aims consisted of modelling decision-making, evaluating the impacts of past policies, drawing up different future scenarios for landuse and desertification, and developing a support framework for desertification policy, the MEDACTION Project represents, so far, the most policy-oriented research project focusing on desertification at European level. Although the increasing emphasis on human action, as embodied in the MEDACTION project, can hardly be described as an explicit trend in the more applied policy analysis and evaluation that still characterises most policy research today, it may be making research-funding bodies - such as the European Commission - more amenable to the propositions of social science researchers and, hopefully, will eventually translate into more pragmatic applications.

The MEDACTION project was a continuation of the MEDALUS Project mentioned above which consisted of three independent sub-projects that ran from 1991-1999. MEDALUS I-III focused on the interaction between climate, ecology and soils on different scales of field experiments, producing models and indicators of the processes of desertification and identifying environmentally sensitive areas as a basis of mitigation actions (Thornes, 1998b; Kosmas *et al.* 1999c; Imeson and Cammeraat, 2002). It was, however, felt that the work conducted in the MEDALUS projects did not sufficiently explore the human dimension of desertification. More information on the policy contexts and land management decision-making was needed in order to implement the obtained technical and scientific information. There was, therefore, an urgent need for an international research project addressing these dimensions of desertification in Southern Europe, and the MEDACTION project was designed to serve this purpose. It maintained the target areas selected during the MEDALUS projects (see Thornes 1993; Thornes and Mairota 1995; Mairota *et al.* 1998), and consisted of four separate project modules including (1) landuse change scenarios at Mediterranean and regional scales; (2) effects of past policies in the target areas; (3) application of decision-support systems for land management scenarios for policy formulation; and (4) development of a desertification policy support framework. Results presented in this book are based on empirical and theoretical research conducted in Module 2 of the MEDACTION Project (effects of past policies in the target areas), where the effects of implementation of a vast array of policies on landuse and land degradation across the physically, culturally and economically diverse target areas were analysed in detail on the basis of stakeholder interviews and farm surveys (see Chapter 3).

1.4 Nature, networks and interpretative power

In this final section of the introduction, we wish to briefly investigate broader aspects related to the need for an inclusion of social science and thickly textured investigations of desertification processes. Indeed, the recently awakened interest in the human dimensions of desertification highlighted in Section 1.3 can be seen as a reflection of wider trends taking place in policy, as well as academic, discourse. In the theorisation of policy implementation as well as in the field of rural sociology, the methodological implications of the 'cultural turn' (e.g. Buttel, 2001), shifting focus onto the social context of studied issues, are explicit. They can be seen to have led to the restructuration of much of the conventional analytical and conceptual framework of the enquiry into environmental and policy-related issues.

With an interdisciplinary input, the cultural turn introduced notions of post-modernism and social constructionism to social science investigations (Hannigan, 1995). Aiming to provide new tools to the analysis of power, the cultural turn contests the structure-oriented and deterministic approach of conventional political economy explanations and emphasises the role of actor-level agency (Buttel, 2001). These trends have been particularly strong in the enquiry into rural society. It can even be said that, currently, interpretative and subjectivist approaches dominate the sociology and political economy of agriculture (Buttel, 2001), where actor-orientation and the likes of network and commodity chain approaches are gaining ground (see also Castree and Braun, 1998). While no other discipline is as well suited to focus on the interface of nature and society as that of *human geography*, for which this interface forms the ontological terrain, there is an evident need for new methodological and theoretical approaches in any discipline to addressing these, in actual fact, old and familiar issues.

The cultural turn, social constructionism and interpretative actor-oriented approaches that represent a shift away from traditional 'modern' structuralist ways of understanding society, can be seen to have evolved in parallel with wider political, economic and technological transformations. The globalisation of both information and resources and technological advances, particularly in relation to how we control and utilise natural resources, not only produce new scenarios of risk, but also change the relationship we nurture with what we call 'nature' (Beck, 1992; Castree and Braun, 1998). In this context, the 'modern era' of the 20th century may be referred to as an era of capitalist accumulation and economic growth, combined in the dominating goal of gaining control in order to make natural resources subservient to these aims (Eder, 1996). More recent technological developments, however, have complicated this relationship with nature, by granting humans power over it to the extent where we have access to the very makings of nature, our genetic make-up. With these kinds of issues, the ethical, value-based and interpretative dilemmas become crucial. We no longer aim to control nature merely to gear it towards modernist production process, but we are also dealing with processes which serve much more intricate interests and could acquire much more drastic autonomy than the technological treadmill referred to by Max Weber. The so far self-evident status of technological development is beginning to be questioned, and at the same time people are asking what is 'nature', what is 'natural' and, in the context of our book, what is 'desertification'?

1. Introduction

It can be argued that one characterising feature of a post-modern society would be that institutions personifying power structures of traditional and modern societies, ranging from that of family and marriage to those of science and technology, are being questioned to an increasing extent (Pescosolido and Rubin, 2000). The uncertainties created by the demise of traditional and modern hierarchies have led to a variety of 'legitimate' values and interpretations, and hence to a situation where power is no longer manifest in institutions, but rather represented by a process of negotiation between competing interpretations (Latour, 1993). The networks of social connections maintain crucial importance in organising society, but are less firm and contain less of a predetermined power structure (Pescosolido and Rubin, 2000). Consequently, the increasing emphasis placed on human perception, both as an organising and/or a constituting factor in making sense of our environment (both society and nature), means that the traditional tools offered by political economy and other structural approaches are no longer sufficient to understand the dynamics of society. Particularly when approaching environmental issues such as desertification, which are located on the interface of society and nature, more attention, therefore, needs to be paid to human action in a space-specific context.

As we are in this book focusing on the policy framework surrounding the issue of desertification in Southern Europe, we will be focusing on the perceptions and aspirations of policy stakeholders - policy designers and implementers, land managers, and the local inhabitants of our four target areas acting both as observers and participants in these processes. We have chosen the *Actor-Network Theory* (ANT) approach (Callon, 1986; Law and Hassard, 1998) as a means to involve the local natural resources and desertification problems in the network of interaction through which interpretative power is defined in the specific policy networks. Part III of this book will focus specifically on interpreting our empirical case study results through the lens of ANT. Through the theory of sociological translation (Law, 1992), ANT allows us to situate and contextualise the place-specific aspirations and local resources (natural, technological and financial, for example) which dictate the severity of the desertification process into a common frame, explaining how the relevant policies are interpreted, and how they simultaneously condition the desertification process.

This approach, therefore, understands social relations as no longer just those existing between people, but as involving both human and non-human actors. Nature is seen to intervene in social processes, without considering natural laws as transcending those of society but in fact participating in the formation of, and subsequently being subjected to, social 'laws' of cause and consequence. These hybrid nature-cultures consist of intertwining interpretative networks of nature, culture, science and technology (Castree and Braun, 1998). Such an approach reveals the organising processes of our society and enables admission of responsibility for how our inevitable interventions with nature proceed, with what consequences, and to whose benefit. This book will, therefore, offer tools to the understanding, and perhaps contesting, of particular production processes, policy and training networks, as well as individual policies related to desertification problems in Southern Europe.

2. Desertification and policies: the global, European and national arenas

J.J. Oñate, M. Juntti and G.A. Wilson

2.1 Introduction: linking policies and desertification in Southern Europe

Desertification, in its many definitions and manifestations, is undeniably a process to which both natural and human factors give rise. As Van der Leeuw (1999) points out, the dynamics of the processes driving desertification operate on various different time scales ranging from thousands of years to weeks, days or even hours. It is obvious that the considered time span has significant influence on which forces appear meaningful; several disappear from sight in the short run, while taking effect over a longer period of time. In the context of Southern Europe, this book explores the policy-related factors and processes that, over an approximately twenty-year period, have contributed to what at present is termed desertification, although recognising the still visible long-lasting effects of other factors.

There are several socio-economic processes, within and outside of the 20-year time span of analysis conducted here, that are in one way or another linked to desertification. Ranging from urban migration and consequent rural depopulation to irrigated agriculture expansion and consequent rise of living standards, these processes form the social dynamics of desertification and have more often than not been supported, if not initiated, by governmental intervention. Some have undoubtedly come about as unintended consequences of regulations and policies, but, nevertheless, have played a significant role in the changes that have taken place in an already ecologically stressed natural environment (Puigdefábregas, 1995). These processes can be seen as key driving forces of desertification, as further discussed in Part II of this book. Viewing the role of policies related to these driving forces of desertification in this context depicts desertification mitigation as a controversial task for national governments. These governments have to design and implement measures to *counter* the effects of past and present initiatives and decisions taken by both the EU and national governments that have often aimed at enhancing economic development thereby *contributing* to desertification (UNCOD, 1977).

The issue of desertification is highly politicised and open to interpretation. Controversies are not limited to the global arena or to the level of national governments. In practice, desertification often acquires status as a recognised problem when rising living standards and associated intensified landuses have led to conflicts over natural resources, the most prominent being conflicts over water (Thornes, 1996; Troeh *et al.*, 2004). For example, the decisions taken by individual farmers to shift from crops traditionally grown in the Mediterranean lowlands - like cereals and permanent dryland crops - to irrigated crops, which at present are common in these areas, has resulted in a vast increase in expenditure of water in agriculture. Further, land abandonment as an alternative to these intensive and often conflict-prone forms of land management can also be linked to deterioration of natural resources in the Mediterranean environment (Margaris *et al.*, 1996; Van der Leeuw, 1999). In this sense, continuation of non-intensive farming and the consequent use of small-scale soil conservation management techniques, such as terracing and gully stabilisation, is beneficial as it can prevent large-scale and much more hazardous and uncontrolled erosion processes that would take place in an unmanaged environment in the Mediterranean climate.

Understanding the political and socio-economic dynamics behind these different forms of human intervention in nature can be seen as essential for managing and mitigating desertification, which, ultimately, is a question of defining the most appropriate uses for natural resources in both an environmental and socio-economic sense (Van der Leeuw, 1999). It is, therefore, crucial to define why certain land management decisions are made. For example, why have many farmers shifted from cereals to horticulture in Mediterranean lowlands? Further, who are the authorities responsible for inducing or accommodating such shifts, and what are the agendas guiding stakeholders' decisions in order to achieve effective changes? The aim of the UNCCD and the subsequent NAPs is to inspire such coordinated efforts to mitigate desertification at national and sub-national level. As further discussed below, the considerable challenge is to achieve the integration of desertification mitigation goals into a vast array of policies guiding land management and the use of natural resources in affected areas. Part II of this book will analyse cases of socio-economic dynamics of desertification with particular emphasis on the policy context and stakeholder networks of land management in four areas in Southern Europe with particularly high propensity to desertification.

This chapter first discusses the emergence of global level agreements on the need to address the problem and the consequent drafting and ratification of the UNCCD, and then briefly addresses the general institutional and political context in which the convention is at present implemented in Southern Europe (Section 2.2). Before moving on to outline specific policies related to desertification in Section 2.3, some conceptual issues concerning the identification of relevant policies will be discussed. In addition to policies and programmes implicitly aimed at combating desertification, a wider framework of landuse related policies will be outlined in Section 2.3, focusing on the impacts of these policies on desertification processes. This will set the scene for Part II of this book which will explore how these policy frameworks function (or not) in the form of four detailed case studies from Southern Europe.

2.2 The UNCCD: genesis, objectives and implementation in the Southern European context

Combating desertification in Southern Europe takes place in the political context framed by the *United Nations Convention to Combat Desertification* (UNCCD). Not only does the Convention, in its five annexes, define which countries are affected by desertification, but it also promotes a certain kind of understanding of the phenomenon of desertification and of the required actions - the goals and processes of drawing up and implementing the *National Actions Plans* (NAPs) at the level of nation states and, ultimately, on the ground. It is therefore important to understand the politics of desertification at global level and the resulting objectives set for the NAPs, which can be presumed both to act as a catalyst and to give direction to most of the significant desertification mitigating actions in the countries in question here. In short, the task of desertification mitigation is that of transposing the global objectives to the local level, which is by no means an uncomplicated process as we argue in later chapters, and as is vividly illustrated by the case studies outlined in Part II of this book.

2.2.1 Global desertification policies: compromise or coherent solution?

The process through which desertification became recognised as an environmental problem demonstrates the significance of global-level scientific and political action in raising awareness of, and inspiring responses to, environmental problems. Scientific interest in studying landuse and land cover changes at the global scale was first expressed during the 1972 *Stockholm Conference on the Human Environment*. However, desertification did not begin to be recognised as a problem until 1977, when the *United Nations Conference on Desertification* (UNCOD) was held in Nairobi, Kenya. As a result of UNCOD, a *Plan of Action to Combat Desertification* (PACD) was adopted, the implementation of which was left to national governments, with an overall coordinating role assigned to the *United Nations Environmental Programme* (UNEP). Fifteen years later however, the Executive Director of UNEP had to recognise the failure of the PACD. Although many authors claim that the reasons for failure are complex and related to local socio-economic conditions and inappropriate policy instruments, the lack of interest from external funding agencies and the lack of funds in developing countries themselves were widely hailed as the main reasons for failure. Indeed, desertification control measures had not been sufficiently integrated into socio-developmental programmes, and there was also pronounced absence of involvement of local stakeholders in desertification mitigating actions (UNEP, 1991; Swift, 1996; Thomas, 1997; Sullivan, 2000).

The lack of funding for the PACD and the lack of effective action regarding desertification can, to some extent, be related to the perception by most developed countries of desertification as a *regional problem*, not necessarily warranting global action, although desertification was brought to the forefront of world attention in the 1970s due to the devastating environmental, social and economic impacts of recurrent droughts in Sahelian Africa (Olsson, 1985; Swift, 1996; Thomas, 1997; Bowyer-Bower, 2003). 'Desertification' was used originally to designate the circumstances of expansion of desert-like conditions (including absence of humans) in economically marginal areas such as the African Sahel, and the UNCOD definition of desertification emphasised this perception, where desertification was seen as the "diminution or destruction of the biological potential of the land, which could lead ultimately to the formation of desert-like conditions" (Pérez-Trejo, 1992, 12).

In the 1970s and 1980s, population pressure and resulting over-intensive land management by local populations were widely blamed for the degradation of African environments (Leach and Mearns, 1996; Fairhead and Leach, 1996). By the early 1990s, however, controversial evidence of the nature and causes of desertification in sub-Saharan Africa, together with increasing voices from the ground contesting the simplistic understandings of local-level driving forces of desertification, paved the way for a renewed approach to the issue at global level (Swift, 1996; Sullivan, 2000; Bowyer-Bower, 2003). Further, the global nature of the problem was acknowledged when the shifting understanding of its causes expanded to recognise that larger arid and semi-arid areas of the globe, outside Africa, where also threatened (Dregne *et al.*, 1991). Accordingly, a new definition was adopted by UNEP, in which desertification was understood as "land degradation in arid, semi-arid and dry sub-humid areas, resulting mainly from adverse human impact" (Pérez-Trejo, 1992, 13). The new, arguably vague, concept of 'degradation' and the expanded spatial dimension of the problem - now covering sub-humid areas in addition to desert-like areas - undoubtedly contributed to increased attention to the problem of desertification well beyond African countries. Nevertheless, African states remained the most active in international action against desertification, promoting the recognition of factors like poverty and the need for alternative livelihoods in resolving the problem, while still arguing for the need for increased funds.

During the preparatory process of the *United Nations Conference on Environment and Development* (UNCED) in 1992, popularly known as the Rio Conference, more than 40 African states pressed, under a common position, for a convention to combat desertification as one of the concrete outcomes to be included in Agenda 21 (UN, 1992). This convention would refocus international attention towards African issues, after binding conventions on climate change and biodiversity had been strongly supported by developed countries (particularly the EU). As a result, an entire chapter of Agenda 21 was devoted to the issue of desertification (Chapter 12 'Managing fragile ecosystems: combating desertification and drought'; UN, 1992). Further, a request was made to the UN General Assembly, at its 47th session, to establish an Intergovernmental Negotiating Committee for the establishment of an *International Convention to Combat Desertification* (INCD), with a view to finalising it by June 1994. The following quote illustrates that the extent and gravity of the problem of desertification was increasingly recognised at the global level in the early 1990s:

> "Desertification affects about one sixth of the world's population, 70 per cent of all drylands, amounting to 3.6 billion hectares, and one quarter of the total land area of the world. The most obvious impact of desertification, in addition to widespread poverty, is the degradation of 3.3 billion hectares of the total area of rangeland ...; decline in soil fertility and soil structure on about 47 per cent of the dryland areas...; and the degradation of irrigated cropland ... with a high population density and agricultural potential" (UN, 1992, paragraph 12.2).

After five negotiating sessions over only 13 months, the INCD came up with what came to be termed the 'United Nations Convention to Combat Desertification' (UNCCD) for countries experiencing serious drought and/or desertification, particularly in Africa (UN, 1994). The Convention was adopted in Paris in June 1994, along with resolutions recommending urgent action for Africa, and interim arrangements for the period between adoption of the

Convention and its implementation[2]. The core of the UNCCD is the development of national and sub-regional/regional action programmes/plans (NAPs and RAPs respectively) to combat desertification, which are to be fully integrated into other national policies for sustainable development, and being flexible and modifiable as circumstances change. Reflecting the emphasis placed on local action in Agenda 21, the Convention itself states that these programmes are to be developed by national governments in close cooperation with donors, local populations and non-governmental organisations (NGOs).

Following the first phase of the negotiation process, the adoption of the UNCCD gave path to a 'post-agreement' phase of negotiations, focused on the development and implementation of the Convention. The Convention entered into force in December 1996, three months after it had been ratified by the required 50 UN members and 30 months after its adoption, representing a slightly longer delay than the ratification of the Conventions on Climate Change (22 months) and Biological Conservation (18 months). Up to six sessions of the International Negotiating Committee were held in the interim period between adoption and the first *Conference of the Parties* (COP; the Convention's supreme governing body) in October 1997. Since then, four further sessions of the COP have been held, and pursuant to articles 22 and 26 of the Convention, a first meeting of the *Committee for the Review of the Implementation of the Convention* (CRIC1) took place in November 2002 and a second in September 2003. As of March 2004, 189 countries had ratified the UNCCD.

In comparison to previous international initiatives regarding desertification, the UNCCD is notable for its innovative approach regarding five specific aspects (see below), most of which are at the centre of the paradigm of sustainable development. However, reflecting the controversial politics behind defining and addressing desertification at the global scale, the road to these new and innovative aspects has been difficult and lined with hard negotiations between parties, often resulting in compromises (Sullivan, 2000; Bowyer-Bower, 2003). Underlying most differences, financial issues continue to be the key element of confrontation between the donor community and affected developing countries, although critics claim that, with closer examination, the problematic issues are much more complex and related to questions of power and interests in land management, and allocation and amount of international funds (Sullivan, 2000).

As the first novel aspect to addressing desertification, the Convention acknowledges that desertification and drought are *problems of global dimensions*, affecting all regions of the world and requiring joint action of the international community to solve them. In spite of wide scientific background supporting this new approach to the problem, its recognition in the UNCCD wording was one major point of divergence between developing and developed countries during negotiations, with the latter fearing the specific connotations that the term 'global' had within the Climate Change Convention. At the Intergovernmental Negotiating

---

[2] The UNCCD Secretariat maintains a web page (http://www.unccd.int/main.php) where exhaustive information on the UNCCD itself and its implementation can be found. The 40-article UNCCD is divided into five sections: Introduction, general provisions, action programmes, scientific and technical cooperation and supporting measures, institutions, and procedures. In addition, the Convention contains four regional implementation annexes for Africa, Latin America and the Caribbean, Asia, and the Northern Mediterranean, which have equal legal status in the Convention. In December 2000, a fifth regional annex for Central and Eastern European countries was added.

Committee on Desertification, the developed countries were, therefore, careful to avoid similar responsibilities for action as established in the Climate Convention, which would have meant considerable alterations to their obligations for assistance.

The second novel aspect, *partnership and partnership building* as called for in the Convention, appears as one of its most significant accomplishments. The UNCCD invites three forms of coordination in the idea that combating desertification on a small scale will be insufficient. These three types of partnerships are to take place between the countries in the South, between developing and developed countries, and with other conventions. Partnership among developing countries is essential to increase aid coordination on the ground, ensuring that good pilot programmes are developed and replicated, and that technology transfer is redirected according to demand indicated at local level. However, Sullivan (2000) argues that developing countries may find it difficult to coordinate their work, not only because of conflicts of interests between different groups, but also because of the traditionally 'top-down' coordination among donors in the North. Partnership building implies that aid flows should be monitored and better assessed, and efforts should be stimulated towards not only funding but also increasing awareness regarding the new approach within international financial institutions.

A third innovative aspect is the call for *involvement of affected local stakeholders and NGOs* in the development of national and sub-regional/regional action programmes to combat desertification (Thomas, 1997). The acceptance of the importance of a participatory 'bottom-up' approach in the development of such action programmes, and the recognition that this element is a pre-condition to successful results, establishes a clear link between the UNCCD and the efforts of achieving sustainable development enshrined in Agenda 21. Moreover, the Convention specifically assigns a role to NGOs in participation. Twenty years ago, few countries would have even considered the usefulness, let alone the political significance and good sense, of NGO involvement. Now, NGOs not only have fruitfully participated in the negotiating process and influenced decision-making, but they are also called upon to play a key role in the development and implementation of the NAPs. As a result, the whole process has been an important capacity-building exercise for NGOs, whose impact on the UNCCD, and whose foreseeable role in its implementation, has set an important precedent for other conventions.

A fourth aspect relates to the fact that the North-South divide manifested itself also in the *scope of the Convention*, which did not provide new financial resources for combating desertification in affected countries. African and other developing countries came to the negotiating table hoping that the UNCCD would provide new and additional financial resources and technical assistance to deal with the causes and effects of desertification and drought. They were particularly interested in addressing the socio-economic causes of desertification, in the belief that it would be impossible to combat desertification unless issues such as external debt, international market conditions, exchange rate variations, pricing, trade policies and poverty were adequately dealt with (Wilson and Bryant, 1997). Developed countries, on the other hand, came to the table with the firmly held position that new and additional resources would not be forthcoming and, instead, called for existing resources to be used more efficiently (Bowyer-Bower, 2003). Moreover, advanced economies preferred to limit the scope of the UNCCD on physical rather than socio-economic causes of desertification. The outcome of this discussion represented a compromise between the divergent positions. On the one hand, no commitment for new and additional financial resources was included in the UNCCD - an aspect deemed to be one of its biggest shortcomings according to many developing countries. Instead, a new awareness on the need to coordinate action and aid programmes was emphasised in the text, addressing the issue of re-orienting development aid. On the other hand, and as outlined above, in the search for explanations and solutions for desertification the Convention has recognised the socio-economic aspects of desertification on top of physical and biological causes. This recognition represents a key change in the attitudes within the donor community, now stressing the need for incorporating these factors into action programmes. Arguably, this allows the UNCCD to be considered as the first 'sustainable development convention' to be negotiated after the Rio Conference.

As for actions taken to re-orient aid flows, until 1994 funding for desertification mitigation rested on bilateral agreements between donors and individual developing countries, a system that proved insufficient to cope with full participation of the latter in INCD sessions. A new 'global mechanism' was established by the Convention in 1994 in order to increase the effectiveness and efficiency of existing financial mechanisms. After long discussions and negotiations, it was agreed that the global mechanism should aim to mobilise, guide and channel resources from existing bilateral and multilateral sources, including countries and international financial institutions, to activities, programmes and projects to combat desertification, thereby increasing aid coordination on the ground. In this sense, the UNCCD introduced the idea of financing the transition to sustainable development, placing the issue of 'coordination' at the centre of the strategy. Administered by the *International Fund for Agricultural Development* since 2000, the global mechanism has played an important lobbying and facilitating role, catalysing the mobilisation of resources from individual countries and existing financial institutions, and encouraging partnership between and within donors and developing countries. However, support from donors has been sporadic, dependent on the practical demonstration by affected countries of their political will within the provisions of UNCCD. In turn, lack of precise funds has been a factor discouraging developing countries from adopting important institutional, economic and social reforms needed to meet the requirements of the UNCCD.

However, as the issue of combating desertification has been re-oriented under the paradigm of sustainable development, an important shift in donor countries has taken place regarding financing of UNCCD implementation. The World Summit on Sustainable Development, held in Johannesburg (South Africa) in September 2002, was a pivotal meeting with regard to these developments. As a result, in October 2002 'land degradation' was added as a new focal area for the *Global Environmental Facility* (GEF), the financial mechanism established in 1991 for international environmental agreements such as climatic change and biodiversity. Implemented jointly by the World Bank and the *United Nations Development and Environment Programmes*, the GEF brings together 173 member governments - working in partnership with the private sector, NGOs, and international institutions - to address complex environmental issues, while supporting national sustainable development initiatives. The decision to make the GEF the financial mechanism of the UNCCD will not only raise the profile of desertification as a major issue, but will also allow countries to access new resources for implementing anti-desertification projects beyond those channelled via the above-mentioned Global Mechanism.

A fifth aspect of the Convention relates to *scientific and technical cooperation* issues, which also represent an innovative approach in the UNCCD. The idea to have a *Committee on Science and Technology* (CST) was driven largely by the presence of similar bodies in the Climate Change and Biodiversity Conventions. But the complex interface between the social and scientific causes of desertification required a unique and innovative approach to determine the character, composition and functions of the CST. After lengthy discussions, the CST was established as a subsidiary body of the COP, integrated by government representatives competent in the multi-disciplinary fields of expertise relevant to combat desertification and mitigating the effects of drought, and open to the participation of all parties. The functions of the CST include advisory functions, data and information functions, research and review functions, as well as functions related to technology, networking of institutions, agencies and bodies. The bureau of the CST is composed to ensure geographical distribution and adequate representation of affected country parties, is responsible for the follow-up of the work of the CST between sessions, and may benefit from assistance of the ad hoc panels established by the COP. However, the requirement of equal representation means that the number of members of the CST is (too) high, and the resulting quick rotation of bureau members is a key factor inhibiting the CST's work and causing inefficiency (Cornet, 2002).

These five innovative dimensions can be seen to make the UNCCD a unique, global, environmental agreement and agenda for action. However, throughout negotiations, developing countries have always expressed their concerns that countries belonging to the *Organisation for Economic Cooperation and Development* (OECD) were trying to downgrade the UNCCD to a lower status than the Conventions on Climate Change and Biodiversity. During pre-agreement discussions, African countries, in particular, had insisted that the Secretariat of the Convention should be an *independent body* outside any existing UN body. However, this option was not seen as financially viable by developed countries, who argued for better use of existing institutions, and a simple facilitative role for the Secretariat. Finally, in January 1999 the Permanent Secretariat was established by the United Nations in Bonn (Germany) to assist the Intergovernmental Negotiating Committee on Desertification and the COP. Yet, partnership among the Rio conventions still continues to be a matter of discussion. Since the three environmental conventions emerging from the Rio Conference have significant overlaps and with processes related to climate change and desertification being particularly difficult to disentangle, combining resources and institutions would make sense. One step towards this direction is that the NAPs will only receive funding from the GEF on the condition that they somehow relate to actions concerning other conventions (Cornet, 2002).

The three conventions (climate change, biodiversity, desertification) have clearly much to contribute to each other. The Conventions on Climate Change and Biodiversity, for example, could benefit from the above-mentioned innovative aspects of the UNCCD, especially the more participatory grassroots approach aimed at dealing with fundamental concerns of humankind such as survival and freedom. However, at present, this may only be possible at the level of lessons learnt from past mistakes, as it remains as yet doubtful how well the UNCCD actually manages to address these issues in practice at local level (Adger *et al.*, 2001; Cornet, 2002; see also Part II of this book)[3]. One factor adding to the doubts concerning actual grassroots involvement is that, although significant emphasis was placed on improving the participation of local stakeholders in the UNCCD process, local voices were rarely heard at the CRIC1 meeting. Even in terms of NGOs, the voice and interface of grassroots communities remained insignificant, with only 46 NGOs present and forming only a small fraction of the 650 stakeholder groups accredited with observer status to the COP (CRIC1, 2003). Further, evidence of a persisting need to tackle the local socio-economic context of desertification was also provided by the case studies reported from affected country parties, which clearly demonstrated the relationship between desertification and eradicating poverty. In addition, it was also suggested that the participatory process of implementation of action programmes has been a more demanding process than initially thought and that it fell short of attracting sufficient resources. In this sense, topics at the root of the problem such as land tenure, agriculture, pastoralism, poverty and food security, deserve further attention before implementation structures are put in place, based on issues such as technology availability, resource mobilisation, and legislative and institutional frameworks. Perhaps resulting from the lack of grassroots representation, issues concerning developed countries, such as resource mobilisation and coordination, were not fully addressed at the CRIC1 meeting, avoiding significant discussions on the root causes of the problems such as the impacts of agricultural subsidies on desertification processes in developed countries - one of the key justifications for writing this book. Thus, the report of the CRIC meeting highlighted that it could be questioned whether the Committee attempted a thorough review of progress made in implementing the UNCCD, or whether it was just a workshop to exchange information and report on 'success stories' (CRIC1, 2003).

The importance of the UNCCD is certainly highlighted by the number of countries who have ratified it - a number twice as high as for the Conventions on Climate Change and Biodiversity. However, resource shortage appears to be a real problem. The CRIC1 report (2003) points out several factors, including an overall decline in official development assistance over the past decade. As an illustration, only around two per cent of the amount of bilateral official development assistance from OECD countries was allocated to combating desertification between 1998-2000 (OECD, 2002). This is clearly a low figure even in comparison to biodiversity-related aid (3%) and climate change-related aid (7%). Ideally, the UNCCD would have much to offer to the other conventions, especially if it were able to find the right balance between international, national and local action and environmental and development objectives. Such coordination would not only result in resource efficiency but also, hopefully, stimulate tangible action. Needless to say, at the moment the modalities for such coordination are topics of further discussions. Indeed, this book aims to make a contribution to these debates by shedding more light on the impact of policies and policy processes on desertification at a Southern European scale.

In conclusion to this section, perhaps the most important accomplishment of all the negotiations that have accompanied the birth and implementation of the UNCCD, is the

---

[3] As mentioned earlier, so far only a first meeting of the CRIC1 has taken place in November 2002 (CRIC1, 2003).

international attention that has been mobilised around the problem of desertification. It is evident that, regardless of the specificities in the text of the Convention, the process itself has been a success in providing a networking forum for those affected by desertification including donors, affected developing countries, United Nations (UN) agencies, intergovernmental organisations and NGOs. These numerous contacts have already laid the groundwork for future partnership arrangements to combat desertification.

Nevertheless, political and participative dimensions still need particular consideration. First, at the political level awareness mobilisation has not yet taken place without conflict. As mentioned, desertification was considered almost exclusively an African problem at the beginning of the process, and African countries were the most interested in including desertification in Agenda 21 outcomes. Indeed, desertification has its greatest impacts in Africa, where combating desertification and promoting development are virtually one and the same due to the social and economic importance of natural resources and agriculture (UN, 1992; UNCCD, 1994; Bowyer-Bower, 2003). However, since the first session of the INCD one of the most contentious issues among developing countries has been precisely the regional implementation annex for Africa. Certain Asian and Latin American governments, while supporting the need for priority treatment for Africa, believed that similar instruments for other regions should be negotiated simultaneously. The possibility that the Convention would focus on development assistance, catalysing the provision of new and additional resources, stimulated a sudden interest in desertification by other developing countries outside Africa, while Africans feared that attention diverted towards other regions would have the effect of diluting the priority they had been guaranteed. The INCD was able to come up with a compromise on this issue by, on the one hand, incorporating four regional annexes on the implementation of the UNCCD, namely those for Africa, Asia, Latin America and the Caribbean, and, the area under focus in this book, the Northern Mediterranean (in December 2000 a fifth annex for Central and Eastern European Countries was added). On the other hand, the UNCCD still gives priority to Africa, both in formal terms with the addenda in its title 'particularly in Africa' (UN, 1994), and in practical terms by having simultaneously passed a resolution entitled 'Urgent measures for Africa'. This resolution entered into force in June 1994, and awareness-rising activities extended all over the region after developed countries began to demonstrate their support. Benefiting from this experience, the annex for Africa is the most detailed and thorough of the regional annexes to the Convention. As a result, by early 2003 24 African countries finalised, validated and adopted their NAPs, four Sub-Regional Action Programmes were finalised and a RAP is also being developed (UNCCD, 2001). It can be argued that the African experience should serve as an example for the other UNCCD Annex regions when preparing their action plans - an issue we address in Part II of this book.

Second, at the participative level real work still lies ahead in translating this political interest to the grassroots level. While developed countries need to commit themselves to implement the UNCCD - entering into productive partnerships with affected countries and increasing their efforts including capacity-building and financial support - one key challenge of the Affected Party governments is that many do not have a tradition of participation. Access to information is limited, and bottom-up input is practically non-existent, with desertification often being the result of flawed government policies and projects (Pamo, 1998). Appropriate legal, political, economic, financial, and social measures need to be adopted in these countries in order to encourage participation (Van Rooyen, 1998; Seely, 1998). One crucial aspect is disseminating information to grassroots and community organisations to ensure that bottom-up input is received and incorporated into action programmes.

While these aspects of desertification mitigation have been studied in detail in the circumstances of developing countries, the participatory aspects of desertification mitigation in Southern European countries remains a weakly explored issue. The following section investigates the particularities of the Northern Mediterranean Annex to the UNCCD and the state and context of desertification mitigation in these Southern European countries (which will also be the focus of detailed analysis in Chapters 4-7 of this book).

2.2.2 Implementation of the UNCCD in Southern Europe

The regional implementation Annex IV of the UNCCD specifically concerns Southern Europe (the 'northern Mediterranean' in UNCCD terms), and designates the four EU Member States of Portugal, Spain, Italy and Greece as Affected Country Parties (UN, 1994). Similarly to the other annexes, the UNCCD identifies particular conditions affecting these countries, including:

a. semi-arid climatic conditions affecting large areas; seasonal droughts; high rainfall variability and sudden and high-intensity rainfall; poor and highly erodible soils prone to develop surface crusts;
b. uneven relief with steep slopes and diversified landscapes;
c. extensive forest losses due to frequent wild and anthropogenic fires;
d. crisis conditions in traditional agriculture with associated land abandonment and deterioration of traditional soil and water conservation measures;
e. unsustainable exploitation of water resources leading to serious environmental damage, including chemical pollution, salinisation and exhaustion of aquifers; and
f. concentration of economic activity in coastal areas as a result of urban growth, industrial activities, tourism and irrigated agriculture.

Annex IV also obliges the Affected Parties to prepare NAPs as a central and integral part of the UNCCD strategic planning framework for sustainable development. In preparing and implementing the NAPs, it prescribes the undertaking of a consultative and participatory process, involving appropriate levels of government, local communities and NGOs (UN, 1994). According to Article 5 of Annex IV, each affected Country Party shall:

a. designate appropriate bodies responsible for the preparation, coordination and implementation of its programme;
b. involve affected populations, including local communities, in the elaboration, coordination and implementation of the programme through a locally-driven consultative process, with the cooperation of local authorities and relevant non-governmental organisations;

c. survey the state of the environment in affected areas to assess the causes and consequences of desertification and to determine priority areas for action;

d. evaluate, with the participation of affected populations, past and current programmes in order to design a strategy and elaborate activities in the action programme;

e. prepare technical and financial programmes based on the information gained through the activities in subparagraphs (a) to (d); and

f. develop and utilise procedures and benchmarks for monitoring and evaluating the implementation of the programme.

In June 1995, representatives of countries, intergovernmental organisations, the Commission of the European Communities, NGOs and other institutions involved in combating desertification met in Almeria (Spain) to hold the First Regional Conference on Desertification in the Northern Mediterranean. The Conference recommended that the Affected Parties should start immediately with the preparatory activities for the implementation of the UNCCD, recognising that the level of awareness and political willingness of Southern European countries towards desertification should be improved in order to launch the preparation of the NAPs (RCDNM, 1995). However, the pace of implementation of the Convention in Southern Europe has been slow, with Italy and Portugal delaying the submission of their NAPs until 2000 and Greece until 2001, while Spain only submitted their NAP in 2003.

Paradoxically, this slow implementation is taking place in countries with long traditions in fighting drought and soil erosion. Therefore, before analysing implementation of the UNCCD framework in the affected countries, and as a preparation for the case studies from each of the Southern European Affected Parties in Part II of this book, it is worth examining the circumstances behind this delay in more detail. In order to illustrate some of the typical problems associated with the development of NAPs in Southern European countries, we will investigate the case of Spain, which sheds light on key constraints and opportunities with regard to (non)-implementation of desertification mitigation initiatives.

Transposing the UNCCD from the global to the local: the case of Spain

As mentioned in the previous section, in the beginning of the 1990s the official definition of the term desertification changed its emphasis from 'expansion of desert-like conditions' to 'land degradation in arid, semi-arid and dry sub-humid areas'. This shift in emphasis led to a more transparent definition acknowledging the global nature of the problem, with consequent expansion of the spatial dimensions of 'desertification-affected' areas both *into* and *within* Southern Europe. Having traditionally covered only the central and south-eastern regions of Spain, areas affected by desertification in Europe were extended to cover all semi-arid and dry sub-humid areas of the four Mediterranean countries (FAO, 1977; UNEP, 1992). However, the shifting meaning of desertification also contributed to confusion concerning both the concept itself and approaches to the problem, leading to the term 'desertification' being challenged as a scientific concept altogether (Helldén, 1991; Rubio 1995; Glenn *et al.* 1998). It was argued, for example, that the concept of land degradation diverts attention from the issue of increased water deficit, which forms the main characteristic of natural deserts. This means that how degradation is defined varies depending on individual or social criteria and values (Pérez-Trejo, 1992; see also Chapter 9). Moreover, the approach to desertification has diverged, with one strand oriented towards better understanding of the physical processes that lead to land degradation, and another aiming at addressing factors that may cause social and economic problems resulting from desertification (Puigdefábregas, 1995).

The case of Spain is illustrative of the consequences that this changing emphasis in the meaning of desertification has had in the social, scientific and political response to the problem. Spain appeared in the first desertification map of the world as the most affected European country, with more than 25% of its territory classified as being under 'very high' threat of desertification. As early as 1981, and following the guidelines of UNCOD, Spain launched the Project to Fight Desertification in the Mediterranean (Pérez and Barrientos, 1986). The objectives of this project reflected Spain's comprehensive approach to the issue, which makes Spanish efforts to stand as a landmark in the efforts to combat desertification in affected Mediterranean countries. The project aimed to analyse the varied resources and factors involved in desertification processes; to determine means and techniques to fight desertification and to undertake integrated planning of preventive and restoration actions in affected torrential river basins; and to provide education and training and to disseminate the project issues among involved experts and local populations. Consistent with the conceptualisation of 'desertification' at the time, the project restricted its spatial remit to semi-arid south-eastern regions of the country and its thematic scope to the problems of soil erosion as the most important morphological process triggered by desertification in Southern European semi-arid regions (Mensching, 1986). This restricted thematic scope was reflected in both the selection of coordinating body, the Ministry of Agriculture, and the 'hard science'-biased composition of the panel of cooperating institutions only including water authorities and research institutes in the areas of meteorology, hydrology, physical geography, biology, forestry and agronomy - basically the same institutions that had traditionally managed soil erosion and drought issues in Spain.

The growing international attention towards desertification in the 1980s and 1990s prompted a perception of soil erosion as a *distinctly Mediterranean environmental problem* while problems such as water pollution or acid rain were seen as 'typical' for northern and central European countries. This reinforced the efforts towards the 'fight against erosion' in Spain, where desertification continued to be perceived as a purely physical phenomenon (see also Chapter 4). Following Spanish entry into the EC, both political and scientific actors continued to adopt a perspective where desertification and soil erosion were seen as the two faces of the same coin and particular to semi-arid Mediterranean environments (Fantechi and Margaris, 1986; CEC, 1992; Geeson *et al.*, 2002).

The above-mentioned shift of emphasis evident in the new definition of desertification adopted by UNEP after 1991 did not help to re-orient the Spanish approach to the problem, which remained skewed in relation to at least three issues. First, other processes leading to desertification, apart from soil erosion, tended to be overlooked in the political agenda and in research efforts, particularly over-exploitation of water resources, soil salinisation, pollution and the sealing of soils. Further, socio-economic drivers of actual desertification problems, such as intensive agriculture, irrigation, urbanisation and industrialisation, were rarely perceived as contributing towards desertification, while more emphasis was placed on overgrazing, forest fires and deteriorating dryland management practices (Rojo Serrano, 1998). Second, action against desertification was narrowed down to engineering means to stop the 'advance of the desert', involving the prevention of forest fires, the hydrological correction of creeks, the afforestation of slopes after terracing, and the building of dams to counteract the effects of drought (Rojo Serrano, 1998). Third, the indirect link between desertification and degradation implicit in the new UNEP (1991) definition increased pejorative associations with the term 'desert', undoubtedly contributing to the negative generalised perception among society of arid and semi-arid environments based on arbitrary connotations of 'degradation' attributed to these environments (Esteve *et al.*, 1990). Undoubtedly, this under-valuation has its roots in the low economic productivity of these systems under traditional exploitation regimes based on dryland agriculture, which has lately been forced to give way to irrigation expansion and tourism developments. As a result, perceptions and attitudes of Spanish stakeholders at all levels, including local populations, not only failed to change, but became fundamentally confused regarding the actual 'desertification status' of these areas, the identification of actual causes of desertification processes, and, most importantly, regarding the design of participatory solutions to the wider desertification problem - negative processes that we will also identify in other Mediterranean countries in our discussions in Chapters 5-7 of this book.

Insofar as this Spanish storyline is applicable also to the other three Southern European countries explored in Part II of this book, this helps explain why questions of water use, pollution, and the spread of urban and industrial landuses and tourism (points [e] and [f] of the Mediterranean Annex to the UNCCD mentioned above) have been difficult to take into account when designing the NAPs and, arguably, will be even more difficult to incorporate in the future during the process of final UNCCD implementation. As this book will show, in Southern Europe the main factor of desertification is the intensification of agriculture and the expansion of irrigated lands (Puigdefábregas and Mendizabal, 1998). However, as the remaining chapters of this book illustrate, agricultural intensification is only the tip of the iceberg in the complex web of interests and interactions governing land management. Most irrigation developments are supported by large financial and technological investments and accompanied by large population movements (including non-EU immigrants in the case of Spain). Moreover, these irrigation schemes are frequently competing for water with urban-industrial and tourism expansions, mainly in coastal areas. In Spain, under the current model of economic growth, a market-oriented view of water use has tended to advocate a 'market-forces' approach (Pérez-Trejo, 1992), conforming to criteria of economic sustainability, but affected by severe environmental problems linked to soil degradation (EEA, 2000). Up to now in the southern EU member states, social, economic and political structures have been able to defer these impacts in space and time. However, although economically wealthy, these systems appear not easily adaptive to further water supply reductions and/or increased environmental degradation, and solutions to desertification problems, therefore, need to be found very quickly.

UNCCD desertification mitigation initiatives at the wider European level

The Spanish example above illustrates the context in which implementation of NAPs to combat desertification takes place in Southern Europe. As shall be described in this section, three challenges characterise this scenario: first, the need to re-scope desertification issues to cover the wider approach of soil degradation as a complex environmental problem, including, in addition to soil erosion and floods and landslides, processes such as depletion of water resources, soil salinisation, compaction, land sealing, local and diffuse contamination, and decline in organic matter and biodiversity; second, the need to translate this wider approach to the analysis of policy drivers for such soil degradation problems, encompassing not only agricultural policies, but also those related to forestry, water, transport infrastructures, tourism, urban and industrial development, all in the context of pressures from economic growth; and third, the need to further strengthen public awareness regarding desertification issues at all levels from local populations to high level policy stakeholders, in order to induce political willingness to address the problem of desertification, and to ensure that effective participation procedures are put into practice in the diagnosis of the problem and in the search and execution of positive remedial action.

As further discussed below, and illustrated in detail in the case studies in Part II of this book, until now affected Annex IV Country Parties are facing difficulties in ensuring an appropriate institutional and legislative framework, enabling an effective public participatory process, and mobilising financial resources for NAP preparation and implementation (CRIC1, 2003). Institutional measures have been taken by all four countries under investigation in this book with the creation of *National Coordinating Units* placed under the responsibility either of the Ministry of Agriculture (Greece, Portugal) or of the Ministry of Environment (Italy, Spain). These mechanisms seem to be truly intersectoral, comprising representatives of the technical ministries involved, as well as a representative of the Ministry of Foreign Affairs and scientists. These actors have responsibility for communication on all matters relating to the Convention, ensuring coordination between ministries and other bodies (both public and private) concerned with desertification (COP4, 2000). Nevertheless, and as Chapters 4-7 will emphasise, it is still doubtful whether there is political willingness to give official status to the NAPs. In addition to the National Coordinating Units, the four countries have initiated a participatory process in support of the preparation and implementation of the action programmes, encompassing representatives of the administrations and the scientific community and NGOs. But the operation of these participatory processes has not always been as desired, in part because of their novelty and unfamiliar nature, but also because of weak involvement of stakeholders and NGOs and lack of an immediate response from their citizens. Finally, financial support for combating desertification has not been explicitly allocated in any of the four countries, and, instead, a re-direction of public financing under existing funds for activities linked to desertification is expected.

However, CRIC1 (2003) also recognised the relevance that instruments such as the Aarhus Convention on 'Access to Information, Public Participation in Decision-making and Access to Justice in Environmental Issues' (UNECE, 1998) might have in strengthening the participatory process. Furthermore, CRIC1 (2003) emphasised the cross-sectoral nature of the necessary efforts to combat desertification and the important role of supra-national policies, such as the EU's environmental and agricultural policies (see below), for the process of implementing the UNCCD in Southern Europe. This 'Europeanisation' of the desertification problem was already clear in the 1972 *Council of Europe's European Soil Charter* (CoE, 2002, 9), in which European soils were considered as a complex natural resource increasingly threatened with various forms of degradation:

- *"Physical degradation due to urban sprawl, erosion caused by development, transport projects or road construction, various types of mining activities, or destruction or compaction and sealing of surface soil as a result of intensive farming techniques.*
- *Biological degradation caused by sediment formation, acidification, natural salinisation and organic impoverishment of soil.*
- *Pollution caused by acidifying, toxic and chemical substances, particularly heavy metals, radioactive substances, dumping of household, industrial or radioactive waste, use of fertilisers and plant protection products, or spreading of sewage sludge or livestock waste.*
- *Degradation as a result of wind or water erosion or inappropriate farming or forestry practices"*

Further, the European Environment Agency, in a joint message with UNEP, recently focused attention on the status of European soils (EEA, 2000), prompting a discussion on the need for a pan-European policy on soil as a basis for managing European soil resources. In light of case study results discussed in Part II of this book, EEA (2000, 5) critically emphasised that

"in the end it is a matter of people and the interactions they have on the natural resources and the limited space available. The problem calls for new policies, including fair pricing, fiscal policies, and strategic planning concerning the use of land and natural resources. There is resistance from economic interest groups for such measures as it is seen as limiting liberalisation and reducing speculative expectations. This may become the biggest challenge for sustainability".

Although the EU has yet to introduce specific soil protection measures, Community legislation deals with soils indirectly through various directives and regulations in the policy fields of agriculture, environment, regional development, transport and research, as discussed in the case studies from the four Southern European countries in Chapters 4-7 (see also Briassoulis *et al.*, 2003). Conscious of the fact that soil is a vital and largely non-renewable resource increasingly under pressure by a range of human activities, and committed to the objective of sustainable development, the European Commission has recently issued a communication entitled 'Towards a thematic strategy for soil protection' (CEC, 2002a). Desertification is here understood as soil degradation occurring in dry areas, where climatic conditions combine with human activities in diminishing the soil's capacity to carry out its functions. The strategy for soil protection considers desertification together with other forms of threats to soils, which, although not applying evenly across Europe, are explicitly linked to a worsening environmental situation that calls for increased Community action. The communication acknowledges the international initiatives towards soil protection, including the UNCCD and the 'sister' Conventions on Climatic Change and Biodiversity, as well as ongoing efforts by member states including NAP preparation processes in Portugal, Spain, Italy and Greece. Environmental policy, the Common

Agricultural Policy (CAP), regional policy and Structural Funds, transport policy and research policy are all identified in the communication as relevant for *soil protection*. This is a semantically distinct approach from that of UNCCD in the sense that it avoids the pejorative or negative connotations that the term *land degradation* poses, and, thus, may be more effective in increasing awareness among politicians, stakeholders and local populations. At the same time, the new approach considers the potential role of the whole set of policies influencing soil protection from a positive side, that is, exploring their potentially beneficial role towards soil protection instead of just emphasising their negative effects regarding land degradation.

Under this positive approach, the communication reviews the existing soil data gathering systems (surveys, monitoring, networking), and discusses the potential role of individual and combined Community policy instruments in soil protection. In this sense, the European Commission is placing soil alongside water and air as environmental media to be protected for the future. As the communication states:

> "The Commission has taken a pragmatic approach directed in the first instance towards the adjustment of existing policies relevant to soil, taking both a preventative approach through the development of new environmental legislation and an integrational approach for sectoral policies of particular relevance for soil" (CEC, 2002a, 35).

Finally, a work-plan and timetable for building a thematic strategy for soil protection are established, significantly envisaging a new communication on soil erosion, soil organic matter decline and soil contamination, in light of soil erosion in the Mediterranean region in particular. Overall, this initiative of the European Commission could arguably emerge with a new directive on soil protection which will ensure that soil protection, including the threats from desertification, is treated as a major issue to be tackled both within and outside the EU.

2.3 Policy impacts on desertification in Southern Europe: the influence of the policy arena beyond the UNCCD

The discussion in the previous sections has highlighted that desertification issues are embedded in more than just UNCCD policy concerns for combating desertification, and that issues of soil and water conservation, as well as many other policy arenas, will also have at times crucial impacts on desertification processes in both positive and negative ways. The following will broaden our discussion on policies and desertification beyond the level of the UNCCD, and will analyse how the wider policy environment can have severe implications for desertification enhancement, mitigation and management. Section 2.3.1 will approach this issue from a conceptual level, while Sections 2.3.2 and 2.3.3 will analyse in detail existing studies on EU policies that have been shown to have affected land management and desertification in Southern Europe.

2.3.1 Implicit and explicit policy effects on desertification: some conceptual considerations

As the key focus of this book is on understanding the linkages between policies and desertification, we also need to consider the impacts that the policy environment beyond the UNCCD may have on land management decisions and resulting desertification processes. In this context, it is necessary to analyse EU and national regulations that may have implications for the use of natural resources in our Southern European target countries.

Before we analyse in detail the key policies that affect landuse, land degradation and desertification in Southern Europe in Section 2.3.2, we first need to discuss issues related to conceptualising the possible impact that any given policy may have 'on the ground'. This is particularly important in any assessment of the drivers of desertification, as in democratic societies (i.e. member states of the EU) human decision-making is shaped, and often dictated, by public policy processes. Three key conceptual issues need to be addressed in this respect: first, the nature of the link between policies and desertification beyond the UNCCD framework, which can be conceptualised as either *explicit* or *implicit* in the case of policies not *directly* addressing desertification issues or desertification processes; second, the fact that policy formulation does not necessarily lead to policy implementation; and, third, issues related to the dynamic nature of the policy environment, where possible conflicts between impacts of different policies and questions concerning the geographical scale, stakeholder interests and power over implementation arise, thereby often yielding unexpected policy outcomes.

First, it is important to highlight that no desertification policies *per se* exist yet in the EU. As highlighted in Section 2.2, the UNCCD only provides a *framework for policy action* (i.e. due to lack of legislative capacity for enforcement, it is ultimately a non-binding convention), rather than being a regulatory device directly influencing the actions of stakeholders on the ground. The first 'tangible' policy level is, therefore, the national implementation of policies in response to the UNCCD framework (e.g. NAPs) and, in the case of the EU, the supra-national EU policy framework. Yet, even at the supra-national and national EU level of policy-making, no actual desertification policies exist. Instead, and as Section 2.3.2 will discuss in detail, a wide array of policies that at first sight do not seem to be linked to desertification will have repercussions for desertification processes by encouraging or discouraging certain forms of land management. In the absence of a direct policy framework, it is, therefore, important to consider how the impact of such *indirect* (i.e. non-UNCCD) policies on desertification can be assessed. Thus, conceptually, we need to acknowledge the difference between 'explicit' and 'implicit' policies affecting desertification.

'Explicit' policies can be conceptualised as those policies *that have explicitly been designed to address environmental issues related to desertification*, such as air, water or soil pollution, conservation-enhancing landuses, and erosion prevention and correction. Environmental policy (mainly drawn up at EU level, but also to a certain degree at national level) can be broadly categorised as an 'explicit' policy, as it contains explicit aims and measures to prevent and/or mitigate surface and groundwater contamination, air pollution, and impacts from waste disposal, all of which prevent harmful effects on soil and, thus, on desertification. Nature conservation policies, such as wildlife habitat and biodiversity protection, would also be 'explicit' in their objectives in relation to desertification, aiming at guaranteeing the survival of soil-protective vegetation cover, as do many measures related to forest conservation and planting. Further, environmental impact assessment regulations, both at the strategic and project levels, can be considered as 'explicit' in relation to desertification in the sense that soil conservation and erosion prevention are usually included among their particular objectives.

Further measures with explicit relevance for desertification processes are found under the *Rural Development Regulation* (RDR) of the CAP. Agri-environmental measures (in particular Regulation 2078/92) are the most relevant among these, where landholders across the EU are encouraged through various agri-environmental schemes to undertake environmentally-friendly practices with potentially beneficial impacts for soil conservation and desertification mitigation. However, afforestation of agricultural land (Regulation 2080/92) is also a relevant instrument of rural development, with explicit linkages to desertification through erosion prevention and soil protection. Arguably, following Agenda 2000, the RDR is set to gain increasing potential for promoting soil protection, and the new rural development plans include a definition of 'good farming practice' and a set of verifiable standards where soil protection receives considerable attention (Robinson, 2004). As granting of compensatory allowances in Less Favoured Areas (LFAs) of the EU is made conditional to meeting the standards for good farming practice, EU rural development policy can, arguably, be considered as an 'explicit' policy, indirectly addressing desertification (see, in particular, Chapter 7). Even within the continuously productivist first pillar of the CAP, some explicit opportunities for desertification mitigation can bee seen to arise in a number of individual market regimes, such as set-aside in the arable sector and the extensification premium in the beef sector (Lowe *et al.*, 2002). As will be discussed in detail in Part II of this book, whether these opportunities for increasing environmental sustainability are embraced by policy implementers and land managers, remains as yet subject to doubt. Nevertheless, policies described here as explicit can be conceptualised to have largely beneficial impacts on desertification mitigation in Southern Europe.

It is the 'implicit' policies that may be the most important with regard to worsening desertification in Southern Europe. Implicit policies can be conceptualised as *those policies that have not explicitly been designed to address any of the environmental issues related to desertification, but that, by influencing the use of natural (and human) resources, nonetheless, may have severe indirect repercussions for desertification.* Most often these implicit policies influence desertification processes by encouraging landholders to manage their land in ways which have negative effects in terms of erosion, water depletion or pollution and, ultimately, desertification. However, and as we will see throughout this book, the impact of these policies with regard to desertification processes are dependent on various factors, including the characteristics of the natural environment in question, and are, therefore, impossible to define without knowledge of the local geographical and socio-political context (see also Clark and Perez-Trejo, 1995).

'Implicit' policies will, for example, include many national and EU water policies, designed to control the use of water for irrigation, but with potentially negative effects for desertification by encouraging increased landuse intensity with concurrent risks of soil degradation and desertification. Implicit policies will, in particular, include most CAP 'first pillar' agricultural policies, and it is among these policies that the most important policy drivers *enhancing* desertification processes are currently to be found (see Chapters 4-7). This policy framework includes, in particular, all EU schemes related to the organisation of specific agricultural commodity sectors, which are likely to encourage intensification of landuse as well as land abandonment - both leading to increased desertification risk. Arguably, and as an example of the often contradictory effects of EU policies, implicit policies in relation to desertification can also be found under the RDR, particularly among those related to farm modernisation measures. Section II of the book will, therefore, place particular attention to this potentially harmful cluster of water, agricultural and structural policies.

Second, in gauging the range of policies that are relevant for desertification beyond the UNCCD framework, it should be kept in mind that the formulation of a policy by 'policy-makers' at the national or supra-national level (in our case the EU) does not necessarily guarantee its 'implementation' (i.e. putting into practice what has been formulated in the policy document), let alone its effective operation at the grassroots level 'on the ground' (Ward *et al.*, 1995; Wilson and Bryant, 1997). This highlights that the *formulation of a policy does not necessarily result in a specific 'impact'*. Therefore, non-decision-making and the non-formulation and non-implementation of policies may be as important to consider, as are tangible policy implementation and its direct effects on desertification processes. Indeed, the whole process of policy implementation should be seen as a *relatively fuzzy decision-making spectrum* rather than a clear-cut point in time at which a specific decision is being made and put into practice (Sabatier, 1986; Winter, 1990; Jones and Clark, 2001). This calls for the need to identify and explore the involvement of various stakeholders in the policy processes surrounding desertification.

Third, a conceptual point that deserves attention in the analysis of the possible effects of policies on desertification processes relates to the *dynamic nature of the policy implementation process* and the responses by target groups. As the case studies from Southern Europe in Part II of this book will show, the policy/desertification interface is continuously changing over time. This means that some policies - whether explicit or implicit - that may have been 'benign' with regard to encouraging desertification at the time of their inception, may emerge at certain points in time as *negative drivers* of desertification processes - in other words, the significance of a certain policy with regard to desertification may vary within the time span of our analysis. This reinforces the argument made above that it can be very difficult to equate 'policy' with 'impact', and also means that the assessment of policy impacts becomes more and more problematic the more recently a policy has been implemented (i.e. the 'direction' and impact of recently implemented policies is hard to gauge). This is clearly the case with most of the CAP accompanying measures established in 1992, particularly with the Agri-environmental Regulation (Regulation 2078/92). A case in point has been the 20-year set-aside measure as part of this Regulation, which, in some areas, appears to have encouraged expansion of arable land and increased erosion due to lack of management practices (see particularly Chapters 5 and 7). In addition, the Agri-environmental Regulation is difficult to monitor both for compliance and for environmental impact (Brouwer and Lowe, 1998; Buller *et al.*, 2000). The main factor raising concern in this respect is the high degree of discretion granted to member states in deciding how exactly the Regulation is implemented, to which the voluntary nature of farm-level implementation adds.

A further and related complicating factor in estimating environmental impact of any policy is the time lag between policy implementation on the ground and the point at which resulting changes in the condition of the environment become 'detectable' and 'measurable'. As a result, for policies implemented from the mid-1990s onwards (i.e. including many UNCCD-influenced policy initiatives such as the NAPs), for example, the actual impact with regard to desertification may not be apparent yet. While the detection of any physical changes depends on the existence and degree of sophistication of monitoring technology and databases, the reactions of the policy recipients (in most cases farmers and landholders in the context of our study) towards these policies depends on complex economic, political and socio-cultural factors that can change relatively rapidly, easily within the lifespan of a policy. In our discussion in the following chapters it will, therefore, be crucial to maintain a focus on the institutions and interests involved in policy implementation (or non-implementation), as well as on the grassroots level of landuse decision-making in order to gauge possible impacts of policies on the ground.

This brief conceptual analysis highlights that all three key considerations addressed here need to be borne in mind throughout our study. Policy implementation processes need to be scrutinised thoroughly so that we can explain some of the non-implementation and policy failures we are likely to observe in our four case study areas. The issue of 'explicit' and 'implicit' policies needs to be investigated in detail in our case study areas, and particular attention needs to be paid to the implicit - in other words indirect and less obvious - policy environment beyond the UNCCD. This should particularly allow us to leave room for 'surprises', suggesting that it may be impossible to anticipate at the outset which policy drivers will be the most important in any given area, and that some of the policies initially believed to have the least impact may emerge as key drivers for enhancing desertification processes.

The next section outlines the main groups of policies beyond the UNCCD considered relevant to desertification at a European scale, while focusing more closely on those that appear to have specific impacts on desertification in the four case studies that will be discussed in Part II.

2.3.2 Non-UNCCD policies related to desertification in the EU

As Section 2.2 has highlighted, the UNCCD encourages the Affected Country Parties to utilise already existing environmental initiatives, instruments and institutions to combat desertification, and emphasises the importance of public participation in mitigating desertification processes. The task of desertification mitigation at the level of national governments is, therefore, that of coordination and planning, involving identification of the relevant policies and addressing their implementation. The NAPs and the *National Coordination Units* are supposed to play a key role in this coordinating task, but, as discussed above, and judging from the statements of the CRIC1 (2003) document, significant progress remains yet to be made in Southern Europe. The need to take stock of national level legislation that influences landuse processes related to desertification, therefore, still persists.

In the Annex IV countries of Southern Europe, natural resource use is to a large extent regulated by EU policies, implemented by national officials, and the focus in this section will be on outlining the relevant EU policy context, with specific focus on policies affecting desertification in the four case studies discussed in Part II[4]. The vast and varied range of policies with either implicit or explicit links to desertification is by no means a facilitating factor in the implementation of the NAPs. The landuse processes linked to desertification are governed by a number of regulations, the impacts of which need to be identified and addressed in order for the objectives of the UNCCD to be met. The CRIC1 (2003, 27) emphasised the significance of planning processes and budgeting at national level, and further argued that "the need for more coherent legislative codes, policy instruments and strategic frameworks dealing with sustainable land management emerged [as] one of the main challenges and opportunities for the UNCCD process."

The idea of policy coordination and intersectoral collaboration is not uncommon in the field of European environmental policy (Jordan, 2002). Indeed, environmental policy integration is now an integral aspect of policy-making, especially as the inclusion of environmental concerns in processes and decisions of public policy-making that are predominantly charged with issues other than the environment is enshrined as a policy principle in the so-called Cardiff Process (Hertin and Berkhout, 2003). The 6th *Environmental Action Programme* outlines seven thematic strategies which, at EU level, aim to tackle soil protection, protection and conservation of the marine environment, sustainable use of pesticides, air pollution, the urban environment, sustainable use and management of resources and waste recycling. Due to their complexity and intersectoral scope, these strategies involve a wide array of actors and processes (Fantechi *et al.*, 1995; CEC, 2000, 2001). This implies a strong need for coordination across different policy fields and for environmental integration (Briassoulis, in press). Moreover, and as discussed above, so far the *Commission Strategy for Soil Protection* can be seen as the most comprehensive European-level initiative that addresses issues relevant to desertification.

The policy fields that require special attention and coordination in soil protection are outlined in the strategy as comprising environmental, regional, agriculture, transport and research policies (CEC, 2002a). Table 2.1 shows a subset of these policy arenas that have repercussions, either directly or indirectly, for desertification in the four case study areas investigated in this book. In these case studies, there are four main policy areas that require particular attention when addressing desertification: water policy, environmental policy, the Common Agricultural Policy (CAP) and what can now be termed rural development policy, which groups together *Less Favoured Areas* (LFA) payments and structural and environmental measures for agriculture, as well as some forestry measures. Although these policies have been deemed significant on the basis of results acquired from case studies, they also represent results from Southern European rural areas in general (see Chapter 3 on case study

---

[4] As the case studies in Part II of this book illustrate, there are also individual cases of purely national-level regulations which can also influence desertification processes.

Table 2.1: Implicit and explicit policies (selection) affecting landuse changes and desertification processes in the four case study areas in Italy, Greece, Portugal and Spain (to the year 2000) (Source: authors).

Water policy:
2000/60/EEC: Framework for community action in water policy (E,P)
80/68/EEC: Protection of groundwater against pollution (E,P)
1975/82/EEC: Irrigation works in mountainous areas and LFAs (G)
676/91/EEC: Protection of water against nitrate pollution (E,G,P)

E = Spain
P = Portugal
I = Italy
G = Greece

Agricultural policy:
136/66/EEC: Olive oil sector (I,G)
120/67/EEC: Organisation of durum wheat cultivation (I)
1035/72/EEC: Fruit and vegetable sector (E)
2727/75/EEC: Cereal, oilseeds and protein crop sectors (E)
268/75/EEC: LFA Directive (G)
1837/80/EEC: Organisation of sheep and goat sector (G)
856/84/EEC: Milk quota regulation (I)
797/85/EEC: Improving efficiency of agricultural structures (E,I,G,P)
465/86/EEC: Less Favoured Areas (E)
466/86/EEC: Less Favoured Areas (E)
1094/88/EEC: Set-aside (E,I)
3013/89/EEC: Organisation of sheep and goat sector (G)
2092/91/EEC: Organic agriculture; designations of origin of products (E,I,G)
2328/91/EEC: Agricultural investment; improving efficiency of agricultural structures (I,G)
2329/91/EEC: Improving efficiency of agricultural structures (E)
2378/91/EEC: Extensification of production in sensitive areas (E)
1765/92/EEC: Cereal, oilseeds and protein crop sectors (I,P)
1766/92/EEC: Cereal, oilseeds and protein crop sectors (I,P)
1910/92/EEC: Organisation of durum wheat cultivation in Greece (G)
2069/92/EEC: Organisation of the sheep and goat sector (G)
2078/92/EEC: Agri-environmental regulation (E,I,G,P)
2079/92/EEC: Early retirement scheme for farmers (G,P)
2081/92/EEC: Protection of designations of origin of products (G)
3950/92/EEC: Organisation of milk sector (G)
536/93/EEC: Milk quota regulation (I)
2019/93/EEC: Improving economic situation of small islands in Aegean (G)
3072/95/EEC: Compensation payments to cereal crop producers (E,P)
2201/96/EEC: Organisation of fruit and vegetable sectors (incl. almonds) (E,P)
950/97/EEC: Improving efficiency of agricultural structures (E,G)
951/97/EEC: Investment in agriculture; improving processing and marketing of agricultural products (I,G)
2309/97/EEC: Organisation of durum wheat cultivation (I,P)
1638/98/EEC: Organisation of olive oil sector (I,P)
2366/98/EEC: Organisation of olive oil sector (G,P)
2467/98/EEC: Organisation of sheep and goat sector (E, G)
1251/99/EEC: Organisation of cereal sector (E,I,G,P)

Table 2.1: Continued.

1253/99/EEC: Organisation of cereal sector (E,I,G,P)
1254/99/EEC: Organisation of beef and veal sector (I,P)
1255/99/EEC: Organisation of milk and milk products (I,G)
1256/99/EEC: Organisation of milk and milk products (G)
1257/99/EEC: Rural development regulation (E,I,G,P)
1259/99/EEC: Direct payments for specified crops; common rules for direct support schemes (E,I)
1260/99/EEC: Regulation of Structural Funds (G)
1493/99/EEC: Organisation of wine sector (I,G)
2529/01/EEC: Organisation of sheep and goat sector (G)

Environmental and nature conservation policies:
409/79/EEC: Conservation of wild birds (E,G,P)
337/85/EEC: Environmental Impact Assessment (E)
479/86/EEC: Protection of the environment in the Mediterranean basin (E)
43/92/EEC: Conservation of natural habitats and wild fauna and flora (E,G,P)
42/2001/EEC: Environmental Impact Assessment (E)

Forest policy:
797/85/EEC: Afforestation and set-aside (E,I,P)
3229/86/EEC: Protection from forest fires (E)
3528/86/EEC: Protection of forests from air pollution (E,I)
4256/88/EEC: Development and exploitation of forests (E,I,P)
1610/89/EEC: Exploitation of forests in rural areas (E,P)
867/90/EEC: Transformation and commercialisation of forest products (I)
2080/92/EEC: Forestry measures in agriculture (E,I,G,P)
2158/92/EEC: Protection from forest fires (E,I)
308/97/EEC: Protection from forest fires (E)

---

methodology and general applicability of results). Below we shall briefly describe the characteristics of these policies, leaving out transport and research policies, which are not featured in our case studies due to their relatively indirect links with desertification processes[5]. We will also outline some of the features of the four policy fields that we argue are crucial in defining the impact of these policies on desertification.

As discussed in Section 2.3.1, it is not just the identification of the exact range of policies, but also economic, structural and cultural forces impacting on their implementation and land management in general, that are fundamental to the understanding and coordination of desertification mitigation. Moreover, when focusing on policy impact, it should be kept in

---

[5] We acknowledge, nonetheless, that transport and research policies, as well as a whole host of other policies (e.g. social policies), may also, at times and in specific geographical contexts, constitute important drivers of desertification (see Briassoulis et al., 2003, and Briassoulis, in press, for further detail).

mind that policies issued at EU level are of a peculiar nature, both in terms of how they have come about and how they are implemented (Jordan, 2002). The European Commission acts as the leading administrative authority in charge of drafting policies in collaboration with committees of experts at central EU level. Although there are slight variations to the process of decision-making depending on policy sectors, it is often a long-winded process, involving various committees, the *European Parliament*, and the *European Council of Ministers*, before the member states finally can implement the legislation. The top-down nature of EU policy-making is criticised by Albromeit (1998), who argues that the under-representation of regional perspectives is endemic in decision-making processes at EU level, where member states are represented by national level authorities. Further, trade-offs and compromises in final policy format are characteristic of the negotiations between the different member states and the Commission in the decision-making process (Wilson and Wilson, 2001; Nugent, 2003). The range of individual policies within the policy areas discussed in this section as having linkages to desertification in Southern Europe is, therefore, wide and varied (see Table 2.1), while the coordination of design and implementation of these policies at EU level, as well as at member state level, is cumbersome and often lacking. The various political interests at play in their design mould the policies themselves, and when it comes to implementation of legislation, the EU is, often due to lack of resources, more or less at the mercy of individual national governments and administrative agencies (Jones and Clark, 2001; Wilson and Wilson, 2001; Nugent, 2003).

2.3.3 Agricultural and environmental policy drivers

Agricultural policies under the first pillar of the CAP are key to our discussion and warrant a more detailed analysis due to their potentially severe impacts on desertification processes in Southern Europe. They provide subsidies for agricultural products, are centrally negotiated, and subject to intense bargaining on behalf of each member state attempting to secure a maximum amount of subsidies for farm income (Wilson and Wilson, 2001). They are implemented and monitored by national agencies, often involving local-level policy officials whose actions the Commission can not always control. In practice, this means that the amount of subsidies a farm receives, and more importantly the area of eligible land or the number of eligible cattle, is ultimately controlled by the discretion of administrative officials below the ministerial level of governance. There are two mechanisms which can be seen to embody this principle of *subsidiarity* at member state level and that have increased in significance following implementation of Agenda 2000. First, optional modulation at national level enables the transfer of funds from direct subsidies of the CAP first pillar towards rural development measures under the CAP second pillar (Falconer and Ward, 2000; Buller, 2001). However, the implications of modulation at farm level are simply the reduction of the amount of money incurred from CAP subsidies, conditional upon farm income, total amount of subsidies received, and use of external labour. A further mechanism with more direct implications for environmental issues and desertification processes at farm level is *environmental cross-compliance*, where the reception of direct subsidies is made conditional upon meeting certain environmental standards (Lowe and Brouwer, 2000). Thus, and in line with the principle of subsidiarity, cross-compliance increases the room for discretion in the implementation of the CAP at national level even further. Although both modulation and cross-compliance can be said to enable the introduction of stricter environmental considerations into the implementation of direct subsidies, the extent to which this actually happens remains questionable (Lowe *et al.*, 2002)[6].

When considering specific Common Market Organisations (CMOs), it is those linked to arable crops, olive oil, fruit and vegetable, sheep and goat meat and beef and veal that have the most significant potential to accelerate desertification processes in the case study areas considered in Part II (see, for example, Chapter 5). Since the 1992 CAP reform, all subsidies paid through the arable regimes have been based on land area. In the cereals sector, this has had clear implications in terms of encouraging farmers to *expand* their arable area. Meanwhile, repercussions in livestock farming are less straightforward. The livestock CMOs contain requirements for stocking density, and the Beef Regime has an extensification premium for which the maximum stocking density has varied from 1 LU[7]/ha to 1.6 LU/ha. However, these limits have proven too high for most environmentally fragile grazing areas (Andersen *et al.*, 2000), often resulting in increases in stocking densities (rather than decreases) with potentially dramatic impacts on desertification processes (see, for example, Chapter 7). Since 1992, the Sheep and Goat Meat Regime has been based on a quota system defining the number of heads for which each producer is eligible to receive a subsidy. Although these quotas are tradable among the producers, a practice known as 'ring-fencing' sets limits on the accumulation of headage numbers in a limited area (Ashworth and Caraveli, 2000). Extensive livestock grazing systems constitute a significant landuse in valuable and traditional habitats at a European scale, but particularly in Southern European mountain, woodland and scrubland areas. As Chapters 4-7 will show, the main problem appears to be that stocking densities tend to increase, either due to changing farming practices (lack of pasture rotation) or because of subsidies received through the CAP second pillar support mechanisms.

In the case study areas where the empirical material for this book has been collected, the CAP rural development policy is relevant to desertification mainly through its structural measures for agriculture (modernisation of agricultural holdings and LFA support), agri-environmental policies, and forestry measures. Whereas the structural support funds for rural areas are now administered through the RDR, previously these used to be governed by separate regulations and objectives (mainly Objectives 1 and 5a and 5b of the structural funds). The aim of LFA support is to maintain farmers on the land in areas that can be described as 'agriculturally disadvantaged', but where farming is seen to contribute to the maintenance of the countryside, including human populations. LFA support is paid in form of a supplement to various farm support measures that apply throughout the EU, including aids for farm modernisation, aid for collective investment, CMOs for certain livestock and crops, structural measures for agriculture, and agri-environmental support (Dax and Hellegers, 2000). Further, certain livestock quotas have been expanded within LFAs and, even more importantly, these quotas do not prevent livestock densities per ha from rising above the given limit for LFA aid. The support for the modernisation of agricultural holdings, in

---

[6] Environmnetal cross-compliance was a voluntary measure at national level until 2005, but has been made a compulsory requirement in the Agenda 2000 mid-term review. This means that, at the time of writing, a European-wide framework for cross-compliance is beginning to be applied by each member state.
[7] LU = livestock unit

turn, is paid in the form of investment support, support for young farmers, support for early retirement and for enhancing education, bookkeeping and management support (CEC, 1997).

Although both the LFA support and the structural development measures have contained principles of environmental sustainability in the past, environmental considerations have gained prominence following Agenda 2000 reforms and their integration into the new RDR. Their status in the post-Agenda 2000 rural policy field, however, remains somewhat ambiguous, given the varied range of motives behind the reform (Robinson, 2004). The need to ensure continuation of farming while also encouraging other forms of rural livelihoods, together with the pressure to provide justification for agricultural support in the global market forum, can be seen as encouraging the integration of environmental and other quality-oriented criteria into the CAP and European rural development funds, but they may also be seen as ulterior agendas inspiring environmental concerns as a façade (Buller, 2001). However, not only are the specific goals of rural and structural development on the ground left almost entirely to the discretion of national, regional and even local-level policy actors and stakeholders, but also the extent to which these are compliant with principles of environmental sustainability (Nugent, 2003).

The agri-environmental and agro-forestry schemes of the EU, both of which have *explicit* desertification mitigating aims (as discussed above) are similarly problematic. Since the passing of Agenda 2000, both regulations have been located under the umbrella of the RDR, and are hence eligible to receive additional funds from direct CAP subsidies through cross-compliance and modulation. However, given the combination of environmental and farm-income linked motivations behind the birth of these accompanying measures as a result of the 1992 CAP reform, these policies are often criticised for compromising environmental targets and functioning just as an alternative source of income (Buller *et al.*, 2000; Wilson and Hart, 2000). Particularly the agri-environmental regulation is notorious for remaining open to interpretation in the implementation phase, to an extent where environmental goals are prone to intended or unintended compromises (Buller, 2000; Paniagua, 2001; Juntti and Potter, 2002). Administrative agents responsible for implementing these regulations at regional and local levels are in a critical position to determine the interpretation of the goal of these policies into practice. Despite the benefit attached to decentralised and deliberative models of policy implementation, mainly in terms of inspiring greater policy ownership at operative level, this is by no means guaranteed in the case of environmental policies in particular (Jones and Clark, 2001; Juntti and Potter 2002). Further exploration of the success of these policies in mitigating desertification or agri-environmental problems more generally, is, therefore, more than justified and has already received considerable attention in European policy research (e.g. Whitby, 1996; Brouwer and Lowe, 1998; Buller *et al.*, 2000; Primdahl *et al.*, 2003). Another obvious cause for concern is the lack of financial weight attached to these policies, particularly in comparison to CAP first pillar funds, although the modulation of subsidy funds at member state level may act to alleviate this disparity to some extent in the near future. The extent to which possible beneficial effects derived from agri-environmental and agro-forestry schemes in terms of counteracting desertification are offset by other policies *encouraging* intensification and unsuitable management practices is one aspect addressed in the case studies in Part II of this book.

A further key policy arena that will guide the analysis in the case studies is linked to what could be loosely termed EU 'environmental policies' - which, on the whole, usually fall into the *explicit* policy category with regard to desertification mitigation outlined above. In this context, the EU LIFE programme forms a crucial funding instrument filling an important gap between the creators of scientific information on the environment and those managing natural resources at the operative level. It was created in 1992 to support the environmental policy objectives of the EU, with the aim of co-financing environmental initiatives which entail both the demonstration and development of environmental measures and instruments (CEC, 2002b). The idea is to further the integration of both environmental considerations into community legislation as well as practical development and diffusion of measures to mitigate and manage environmental problems in EU member states and in certain third countries (mainly those bordering on the Mediterranean Sea). The LIFE programme is divided into three parts. 'LIFE Nature' is aimed at supporting the implementation of the Habitats Directive (43/92/EEC), 'LIFE Environment' focuses on environmental innovation and development of management and mitigation practices more generally, and 'LIFE Third Countries' supports the development of environmental legislation and administration in certain less developed countries outside the EU and in new Eastern European EU member states (CEC, 2002b).

As already discussed above, the extent to which desertification is tackled by environmental policies in general has been subject to criticism (e.g. CRIC1, 2003). In light of the history of European environmental policy, concerns appear justifiable (Ward *et al.*, 1995; Jordan, 2002). Initially, EU environmental policy had more of a coordinating than regulating role, including, among others, water policies (also deemed relevant for desertification issues in the case studies of Part II of this book; see Table 2.1 above). The beginnings of environmental policy at European scale were characterised by efforts to ensure that various national environmental standards and regulations did not form barriers to the free movement of goods and services in the area of the then EC (Liefferink *et al.*, 1993). The numerous policies concerning environmental issues are based on aims and objectives defined in *Community Environmental Action* programmes, the first of which was adopted in 1973 (Hildebrand, 2002). Environmental policies, and in particular conservation-oriented policies, face an obvious need to be adaptable to national, regional and local environmental circumstances and values. This has meant that the principle of subsidiarity is central in the implementation of environmental policies (see above discussion on EU agri-environmental policy, for example) which are mostly formulated into directives, binding only in terms of defining objectives to be achieved, whilst leaving national governments free hands to decide over form and method (Hildebrand, 2002). In the case of both regulations and directives on environmental issues, like control of pollution and waste, implementation is characterised by negotiations and trade-offs between national interests which take place throughout the policy process (Jordan, 2002). Nevertheless, EU environmental policy has induced a considerable increase in national level legislation in member states (Ward *et al.*, 1995), although a completely different matter is how, or rather whether, resulting legislation is implemented into practice.

The environmental policies most relevant to desertification in the case study areas of Part II of this book are the *Habitats Directive* (43/92/EEC), which designates certain areas as protected with limited options for management; the Directives on *Environmental Impact*

Assessment and *Strategic Environmental Assessment* (337/85/EEC), which define procedures for assessing the environmental impact of potential projects, plans and programmes; and the *Water Framework Directive* (2000/60/EEC), which ensures the protection of water from pollutants and sets requirements for a river-basin based management system. Here cross-border cooperation between administrative regions and nation states is required in water management and protection, in terms of maintaining good standards of water quality but also in terms of water quantity and defining a need for common principles for controlling water abstractions and impoundment where these may be relevant. Most of the legislation controlling irrigation and setting quotas for water use, however, are established at the level of the nation state (see, in particular, Chapter 4).

2.4 Conclusions

This chapter has provided an analysis of global desertification policies linked to the UNCCD and non-UNCCD policies established at EU or Southern European level that have an impact on desertification processes at the local level. It was highlighted that the UNCCD provides an overarching framework for policy implementation, rather than being a regulatory policy mechanism itself. As a result, the transposition of UNCCD guidelines to the EU and national levels within Europe has, at times, been difficult.

This situation is further complicated by the fact that non-UNCCD policies emerge as at least equally important with regard to both desertification mitigation and enhancement. Our conceptualisation of these policy clusters into either *implicit* or *explicit* policies linked to desertification particularly illustrates the complexity and vast scope of policy dimensions of desertification at European level. While some explicit policies (especially environmental and conservation policies) offer potential for positively managing desertification, our discussion has also highlighted that policies linked to agriculture or water management can increase desertification problems in Southern Europe by encouraging landholders to further *increase production intensity* on already vulnerable land. This means that in the *absence of a holistic desertification policy framework* at EU level, integration of desertification mitigation aims into different policy arenas remains a considerable challenge for the EU and the NAPs - an issue we return to in our concluding Chapter 10. In particular, the process of drawing up the NAPs in the four Annex IV countries under investigation in this book (Spain, Portugal, Italy and Greece) is as yet relatively unexplored and has not substantially advanced to the phase of implementation at the time of writing. Ideally, the NAPs should form coordination frameworks for the implementation of all the policies and projects discussed here.

This chapter will form an important basis to understand case study-specific desertification and policy issues in our four case study areas discussed in Part II of this book. As will be evident from both these case studies and the theoretical analysis of existing and emerging actor networks in the policy framework affecting desertification in Southern Europe explored in Part III, there are severe barriers relating to political institutions, interests, knowledge and communication among the relevant policy stakeholders, that inhibit the functioning of the Southern European NAPs in practice. The remainder of this book will analyse the nature and causes of these barriers in more detail and will discuss how they can be overcome.

Part II

Case studies of policy processes and desertification in Southern Europe

The second part of this book focuses on specific case study material on policy processes and desertification issues. The methodology adopted for the collection of empirical material is explained in **Chapter 3** as a basis for understanding the selection of specific case study areas in Southern Europe. The subsequent chapters will analyse in detail issues of desertification and policies in Spain (**Chapter 4**), Italy (**Chapter 5**), Portugal (**Chapter 6**) and Greece (**Chapter 7**). Each chapter is written by expert teams from these countries that were in charge of collecting and interpreting empirical data related to desertification issues in their case study areas. Each chapter contains a brief description of the case study area, a detailed analysis of local and national desertification issues, and an assessment of the impact of the local, regional, national and international policy environment with regard to exacerbating or alleviating desertification problems.

Empirical information provided in Chapters 4-7 will then form the basis for theoretical analysis based on actor-network theory in Part III of the book, assessing the importance and influence of actor networks operating at the local, regional and national scales in the case study countries, and analysing consequent implications for the various discourses that influence definitions of desertification among stakeholders.

3. Using case studies and actor orientation to explore policy impacts on desertification: conceptual and methodological considerations

M. Juntti

3.1 Introduction: analytical focus and aims

As discussed in Part I, a crucial task of governments designated as Affected Parties in the UNCCD, and a requirement stressed as still incomplete by the CRIC1 (2003), is the identification of the range of policies regulating environmental management practices linked to desertification. Although Chapter 2 gave a brief synopsis of different policy arenas influencing desertification processes in Southern Europe, this alone is still not sufficient in order to fully understand and effectively manage the links between policies, land management practices and desertification. As argued in the introductory chapter, we believe that successful implementation of environmental policy instruments like the UNCCD requires the exploration of the wider socio-economic context - both institutional and interpretative aspects of the desertification issue. We pay detailed attention to the processes of policy implementation and, more precisely, to the range of interests and actors and the negotiation of power among these actors as essential influences on policy outcome and impact on desertification in our four case study areas. This approach stipulates the existence of a *supportive discourse* as an essential component of policy implementation and natural resource management. Moreover, it portrays policy institutions and policy goals themselves as in a state of constant evolution, benefiting from criteria such as clarity of values and concordant moral interpretations, rather than just being subject to a linear process of implementation governed by rigid administrative structures (Lowndes and Wilson, 2003).

These assumptions have significant methodological implications for how the policy links to desertification are addressed in research. The emphasis on interests, interpretations and discourses sets the requirement for a hermeneutic analysis of *thickly textured* empirical material. As a consequence, we employ a bottom-up approach to policy implementation and adopt an understanding of the notion of *power* as vested in the interaction that is taking place in actor networks of policy implementation and environmental management. This is where our approach most clearly relies on ANT, according to which an actor does not have power unless other actors in the network act according to an interpretation of reality promoted by this specific actor (Latour, 1993). With such significance granted to context, we argue, that at this crucial point of beginning of the complex task of unravelling the socio-economic and policy-related causes of desertification in Southern Europe, the case study approach best serves the need for information and understanding. Results from our four case studies from Southern Europe explored in the following chapters will be synthesised in a conceptual analysis in Part III. The conceptual analysis views the 'desertification narratives', collected in the form of stakeholder interviews in our four Southern European case study areas, through the lens of ANT. This approach enables us to identify actor roles and associated power, as well as ruling interpretations manifested in action.

Interpretative power defines the moral grounds on which the goals and practices of policy implementation and natural resource management are defined at national and sub-national levels. The main aim of such an actor-oriented interpretative approach is, therefore, that of uncovering the interpretative and moral premises of land management and policy implementation in our four case studies, and viewing the policy links to desertification in this light. This complements a more descriptive analysis of the structures and processes of policy implementation and involves the local natural resources and desertification problems in the network of interaction through which interpretative power is defined in the specific policy networks. Through the theory of sociological translation (Law, 1992), ANT draws the place-specific structures and aspirations, as well as local resources (natural, technological and financial, for example) which dictate the severity of the desertification process, into a common frame, thereby explaining how relevant policies are interpreted and how they simultaneously condition the desertification process.

Interpretative and actor-oriented context-sensitive approaches are commonly criticised of relativism and contingency, as is inferring general conclusions and concepts from case study material in general. To enhance validity and transferability of results, our analysis is based on a comprehensive consideration of both the physical conditions and economic and administrative structures of our case study areas[8], as well as a combination of quantitative and qualitative data (see below). In particular, our case studies are situated in areas where physical susceptibility to, and extent of, desertification have been defined and mapped by extensive previous research (see, for example, Kosmas *et al.*, 1999c).

Our aim is to develop so called meta-level concepts in terms of policy agendas describing the expressed interests, perceptions, aims and actions that are accommodated and perhaps also supported by the administrative, economic and environmental conditions of our case study areas, as portrayed in the in-depth interviews of policy stakeholders. Relations of power in terms of central and marginal actors and more specific features of the different actor roles will also be extracted from stakeholder accounts. Inferences made on the basis of qualitative information are viewed critically in light of quantitative information concerning the selection of policies that are actually applied at farm level (see Chapter 2) and the composition of the information sources through which farmers are dawn into the policy network.

3.2 The case study methodology

In focusing on the impact of past policies on the processes of desertification taking place in the case study areas, our study combines material from desk-research, in-depth interviews and surveys conducted by four teams of researchers in four case study areas in

---

[8] Our approach, therefore, falls into the broader arena of contextual constructivism, where the prevailing conditions and their history are acknowledged as the framework in which problems and claims are defined (Hannigan, 1995).

Southern Europe. Forming a follow-up project to a range of natural science projects (MEDALUS I-III), the MEDACTION project, upon which this book is based, focuses on the socio-economic dimensions of desertification in four case study areas inherited from the MEDALUS projects: the region of Alentejo in Portugal, the Guadalentín valley in Spain, the Agri basin in Italy, and the island of Lesvos in Greece (Figure 3.1). These four areas were chosen for three key reasons: first, because all have suffered from desertification in the past, and in some areas desertification processes are getting worse; second, an abundance of information on the extent and nature of desertification, encompassing all its physical dimensions, is available from the work conducted in previous MEDALUS projects (see, for example, Geeson *et al.*, 2002); and, third, because individual national research teams had relatively easy access to the areas and their stakeholder networks (i.e. preliminary contacts and information were available).

In-depth interviews of policy stakeholders formed the central data collection method in the four case study areas. Interviews were necessary in order to trace the networks of policy actors, to gain access to the intricate policy processes taking place in the case study areas, and to map the official and unofficial interpretations and agendas of policy implementation and natural resource management, as the only source of this type of information is often policy stakeholders themselves. Thus, qualitative analysis of interview data shed light on land management practices and policy processes relevant to desertification in the case study areas, and provided information on the environmental, socio-economic and administrative contexts as perceived by interviewed stakeholders. The span of analysis of retrospective policy impacts on desertification has been limited to approximately 20 years (i.e. roughly since 1980), mainly due to the limits of human memory as interviewed stakeholders could not be expected to describe phenomena much beyond this time frame. In order to ensure compatibility of the results derived from stakeholder interviews across the four case study areas, a loose structure of data analysis by the national teams was drawn up to guide the collection and preliminary analysis of the qualitative data. However, the detailed structure

Figure 3.1: The four case study regions (Source: authors).

of the interview schedules was left in the hands of the four individual research teams. Thus, the schedules could be tailored to fit the information needs and characteristics of the individual case study areas, but still maintain a common structure and produce comparable information.

The preparation of the stakeholder interviews started with desk research, with the aim to map the stakeholders involved in policy and land management processes linked to desertification. The identified stakeholders represent national (except for the Greek case study area), regional and local level administrators involved in decision-making and policy implementation, experts involved either in research and operative level land management, and actors who hold a stake in desertification through 'consumption' activities by living or holding a business in the case study areas. Policies relevant for affecting desertification in the case study areas (see Chapter 2) also informed the content of the stakeholder interviews. In order to ensure compatibility of the results derived from the stakeholder interviews, a step-by-step analytical structure was drawn up for the national teams to guide the qualitative analysis of the data (Table 3.1).

The normal limitations of this type of methodology had to be considered with regard to the nature and reliability of results derived from research by means of interview data. Qualitative data themselves are social constructs, influenced by researchers' assumptions of social reality, and by the methodology and practices used in the research (Silverman, 1997). The subjective choices and assumptions of the researcher have, therefore, at all times to be explicit. We acknowledge that the interview situation is a social situation promoting certain types of interpretations and ways of displaying knowledge. Factors like mismatching

Table 3.1: Step-by-step analytical structure drawn up for national teams to guide qualitative data analysis (Source: authors).

The stakeholders of desertification
- Identifying actors who contribute to, and sustain, discourses of desertification, as well as those who influence and take part in implementing the goals of land management.
- Identifying connections between stakeholders and discussing the power structure of the networks.

Stakeholder perceptions of desertification
- Outlining the different definitions/discourses of desertification portrayed by stakeholders.

Policy impacts on desertification
- Discussion based on a list of policies compiled through desk research (see Chapter 2), but also including possible policies that were not on the list but that respondents mentioned.
- Focus on specific examples of effects of policies that have worked to mitigate or enhance desertification.
- Differentiating between real and perceived impacts, how realistic respondents' impressions of policy effects are, and how much the power structure of the stakeholder network affects policy impacts in the area.

---

interactional roles of the interviewer and the respondent, difficulties of 'self-presentation', and the fact that the respondent may, for various reasons, wish to lie about or distort their actual thoughts or experiences (the 'tales of self') may lead to misunderstandings between interviewer and interviewee. Building rapport and establishing trust and familiarity were, therefore, necessary for successful interviews to take place in the case study areas. We acknowledge throughout that the social context of the interview is intrinsic to understanding any data that has been obtained, which was the main reason why interviews were conducted by researchers native to the target countries and familiar with their case study areas.

Following the in-depth interviews, *a questionnaire-based farm-level survey* was conducted in each case study area (during 2002-2003) with about 100 respondents in each area (n = >400 in total) to validate results of the stakeholder interviews. Although farmers' views on the impact of past policies on desertification had, to some extent, been explored during stakeholder interviews (mainly through interviews with representatives of farmers' organisations and of individual farmers), much additional information about *policy drivers* and *landuse changes* leading to desertification could be obtained through the questionnaires. Farm questionnaires were the best option for increasing the accuracy of the research results by means of method and source triangulation (Baxter and Eyles, 1997). In all case study areas agriculture forms one of the most prominent landuses and, therefore, holds significance from the point-of-view of desertification. Farmers were, therefore, thought to be an adequate source of information concerning the validity of most of the interpretations drawn from the stakeholder interviews. The questionnaire survey aimed not only at identifying perceptions of desertification and reasons for significant landuse changes, but also to establish what the actual policies guiding landuse at farm level were and what the most significant landuse changes have been at farm level.

The iterative methodology adopted in this project meant that results of the stakeholder interviews could inform the design of the survey questionnaire, both by indicating what phenomena and policies were significant from the point-of-view of our research question, but also by revealing what the qualitative survey could *not* reveal. The questionnaire survey was designed to complement the qualitative interviews on the following points in particular: (a) how individual farmers linked into actor networks and with the help of which sources of information (farmer organisations/advisers/officials) to understand how agendas identified in the stakeholder interviews were transferred into practice; (b) which policies were significant for land management decision-making at farm level; and (c) to discuss policy-related driving forces of desertification identified in the stakeholder interviews. It was agreed that only the general issues that the questionnaires should cover should be pre-defined, while the detailed structure of the questionnaires in individual case study areas was left in the hands of the individual research teams. Thus, the questionnaires could be tailored to fit the information needs and characteristics of the individual areas, but still maintain a common structure and produce comparable information. Issues that were covered by all questionnaires were agreed as follows:
- farm structural information and present landuse practices;
- landuse change and reasons and timing (mainly open-ended questions);

- policy impacts (mostly closed questions supporting and complementing information acquired through qualitative interviews, covering only policies which had been argued to have an impact on landuse by being implemented at farm level);
- interaction with administration, extension and advice services (identification of most important sources of information and advice, particularly in relation to key policies and landuse decisions);
- farmers' policy demands and their willingness to accept different policies (optional).

The use of open-ended, as opposed to closed, questions was left to the discretion of research teams, although open-ended questions were strongly encouraged. The main restrictions for the use of open-ended questions were the limited resources available for data analysis. All teams segmented the farmer sample, so that a smaller sub-sample of particularly interesting cases were subsequently selected for more open-ended questions to provide data that could further complement results from the stakeholder interviews (see Chapters 4-7 for more detail). Open-ended questions concerned mainly farmers' perceptions of desertification, both in general and on their own farm, and reasons for landuse changes that had taken place in the last 20 years.

A common sampling strategy was also established for the farm surveys using the following three criteria. First, the desertification classification developed in the MEDALUS project (see Kosmas *et al.*, 1999c) was used as a basis for sampling, so that farms in the case study areas classified as 'very severely' or 'severely' desertified could be included (see, for example, Figure 6.2 in Chapter 6). In addition, each research team used available national and regional databases to identify a sufficiently representative selection of farms, taking into consideration specific characteristics of the case study area particularly relevant to both desertification and the application of the concerned policies on the basis of stakeholder interviews. Second, sub-samples of farms were selected to explore specific questions in more detail (see above). Third, sampling did not have to be strictly representative, but selected farms should possess characteristics that rendered them relevant to the research questions. For example, small farms were only needed in the sample if farm size was relevant to the way policies were implemented. In the Spanish case study area, for example, both dryland and irrigation farmers were included in the survey sample as both farming areas showed evidence of policy-induced erosion (see Chapter 4 for more detail).

3.3 The UNCCD and the influence of policies and actor networks on desertification in Southern Europe

This final section of the methodology discusses the role of stakeholder interests in policy implementation in general, and, by considering the task of desertification mitigation as defined by the UNCCD, provides further justification for our actor-oriented approach to policy impacts on desertification. It also explains why the subsequent case study discussions in Chapters 4-7 will place specific emphasis on understanding the constraints and opportunities within actor networks for transposing the UNCCD policy framework to their localities.

Given the emphasis on participation and supporting action at ground level, the UNCCD can be interpreted as an effort to stimulate economic and social change through a bottom-up approach by raising awareness of environmental problems, like desertification, and empower local people to manage natural resources in a sustainable way (Cornet, 2002). As Chapter 2 highlighted, the NAPs do, accordingly, spell out the task of managing desertification mitigation as an interdisciplinary issue requiring the participation of a range of actors from various policy sectors and from administrative, grassroots or operative levels of natural resource management. This deviates from the conventional, top-down manner of policy design and implementation in the EU, and while the requirements of ground-level participation and the actual process of drawing up and implementing the NAPs are not clearly defined or controlled, the vast framework of relevant policies (see Chapter 2), and the variety of actors holding a stake in land management, form a complex and challenging ground for achieving any kind of coordinated change, as exemplified by the detailed analyses in each of the four case study areas in the following chapters. While the UNCCD requires that affected countries draw up NAPs, *no formal legal sanctions exist* and strong moral pressure is hoped to be enough to ensure that NAPs are formulated and implemented and that the set targets are met (European Commission, 2000). This ambiguity of both measures and targets further increases the methodological significance of interpretations and the need to understand the mechanisms, processes and strategies that mediate policy goals, power and knowledge at different levels of governmental intervention and desertification mitigation.

Ensuring integration of the goals of the UNCCD is the task of the *National Coordinating Units* (NCUs) within the central administration of individual countries, in order to facilitate collaboration between the different ministries in the implementation of the NAP, and it was compulsory for each country to set up such NCUs. The *National Focal Points* (NFPs) also play a key role in this. However, although the NAPs are now beginning to be drawn up, mainstreaming the UNCCD objectives into government strategies, planning and budgeting is lagging behind (CRIC1, 2003). This problem has also been noted at national level, and discrepancies have been pointed out, particularly in the field of environmental policies which do not fully correspond to requirements set in the UNCCD (COP4, 2000). Better inter-ministerial coordination of cross-cutting policies is still required, as well as collaboration with other relevant conventions.

Inter-policy coordination at the EU, as well as national and regional levels has traditionally been weak, posing a problem to European environmental policy, which has only of late begun to be designed in terms of inter-sectoral strategies, as discussed in Chapter 2. In fact, the implementation of the NAPs fulfils all the criteria for a particularly challenging policy change (Crosby, 1996). The NAPs involve an open policy field of several different sectors and institutions and several stakeholders with different interests, yielding both positive and negative reactions - with important repercussions for the adoption of an actor-oriented methodology used in our study. Moreover, the UNCCD is a type of international initiative that often alienates stakeholders and even governments, particularly if there is a lack of external funding, as is the case with the implementation of the objectives of the NAPs (e.g. COP4, 2000; Cornet, 2002). Mainstreaming the UNCCD objectives into national strategies, plans and policies does not ensure their implementation on the ground, or even the consideration of desertification issues by regional and local governing bodies, as the following chapters will amply show. The realisation of concrete goals at operative level requires, among other issues, legitimising these goals among stakeholders and acquiring the needed resources (Crosby, 1996). Ultimately it is down to individual policy officials and other stakeholders to ensure that NAP objectives are considered in the implementation of relevant policies, and our methodology with its focus on grassroots stakeholders and 'street-level-bureaucrats' is sensitive to this issue. Ensuring that these officials have the motivation and the required skills and knowledge is part of the challenge that the countries participating in the convention are facing. This is why we stipulate the consideration of interests, interpretations and discourses as well as structures and processes of policy implementation as necessary for ensuring sustainability of results in desertification mitigation.

Whether it is a question of policies with implicit or explicit links to desertification (see Chapter 2), unexpected policy outcomes and problems in achieving the desired policy goals are more than likely (Hudson, 1999). In several cases, the policy drivers of desertification are in fact a result of *competing interpretations* of policy officials and other stakeholders concerning policy goals and justifiable ends. As actor-oriented approaches imply, the most significant locus of power in land management is stakeholder interaction that is taking place (or not) in a network of actors (Latour, 1993). We claim that attention should, therefore, be drawn to the processes of *communication* and *interaction* between these stakeholders in both governmental intervention and land management decision-making (see also Moxey *et al.*, 1998; Murdoch, 2000). As Part III of this book will highlight in detail, policy implications can be seen as a result of a process of translating policy goals into practice, usually taking place in intricate networks of decision-makers, policy implementing officials, experts, land managers and various different interest groups whose domain the policy touches on. For example, Williams (2001), who investigated administrative and scientific experts in action in the 1st wave of market reforms in developing countries, highlighted the crucial role of policy technocrats in determining successful policy change, implying the need to consider both formal and informal patterns of governance, including decision-making rules, standard operating procedures and institutional arrangements in ensuring policy success. According to Williams, these factors privilege certain actors over others. Further, as Crosby (1996) emphasised in the context of Lipsky's (1980) theory on street-level bureaucracy, such organisational cultures are often built and institutionalised around a certain understanding of policy goals and clients, which means that policy changes often require changes in organisational cultures and interpretations.

In Part III of this book, we use ANT to explore the range of interpretations competing for dominance in the framework of desertification in Southern Europe. We will also use the ANT approach to understand how, in addition to structural manifestations of power, the loyalties of desertification stakeholders and subsequent representations of natural resources and desertification interlink with certain interpretations of policy goals and target groups and, in turn, impact on the way policy goals evolve and are translated into policy outcomes that themselves influence desertification (Law and Hassard, 1998). In particular, we will identify discourses of desertification, different definitions of the phenomenon, and the implications these hold for required actions and for the allocation of responsibility. We will also extract interpretations of which policies are perceived as relevant for desertification, and how the goals and implementation procedures of these policies are perceived and justified in the interview accounts. These policy agendas and discourses of desertification are exercised in a structural and procedural context, and can be likened to what Jones and Clark (2001) call the modalities of governance - the mechanisms, strategies, procedures and instruments that are used to coordinate power, knowledge and goals in natural resource management in the four case study areas.

Actor orientation does not imply that structural matters will be overlooked in studying policy impact. In Southern Europe, structural deficiencies at national level and the hierarchical nature of EU policy design and implementation, with associated considerable implementation defects between EU institutions and the member state level, are seen as particularly problematic barriers of progress in environmental protection and conservation. These problems are often referred to as the *Mediterranean Syndrome*[9] (La Spina and Sciortino, 1993). Typical symptoms of this syndrome are clientilistic administrative traditions and typically autonomous conduct. One aspect that marks the implementation of several policies and regulations in these countries is the determined decentralisation of administrative power and responsibility that took place in the 1970s and 1980s and which has led to uncoordinated policy action and deterioration of accountability at national and EU levels (Putnam, 1993; Paniagua, 2001; see also Chapters 4 and 5). Another characteristic is the allegedly weak civil society, which makes the promotion of other than economic interests rare or at least difficult at a local level. This weakens the pull factor, for example the force of public opinion exercised through NGOs and voting behaviour, which can be a decisive force in the adoption of sustainable policies and causes in Northern European member states (Pridham and Konstadakopoulos, 1997; Börzel, 2000). Although progress in adopting environmental targets is made at national level - albeit usually in response to external pressures - problems such as fragmentation of responsibilities, limited power of environmental institutions, and lack of policy infrastructure as well as facilities and expert knowledge, are frequent (Pridham and Konstadakopoulos, 1997). Consequently, the Mediterranean member states are often said to lack initiative in environmental policy issues, only acting in response to demands from Brussels, if at all. While acknowledgement of the problem of desertification at national level is only recent and not altogether uncontroversial, grassroots level awareness is still seen as lacking (Mourão, 1998; Tsolakidis, 1998).

However, it should be pointed out that the Mediterranean Syndrome has been criticised as a gross overgeneralisation, which does not take into consideration the variations in both administrative structures and political cultures of Mediterranean countries (Börzel, 2000; Diamandouros and Gunther, 2001). Our interpretative approach to the implementation of policies linked to desertification in both Parts II and III of this book can be seen as a response to this criticism. While some claim that the Mediterranean Syndrome is beginning to give way, and that both environmental NGOs and civil society are beginning to gain power to interfere in policy issues (Börzel, 2000), the four case studies from Greece, Italy, Spain and Portugal, explored in detail in the following, show that structural deficiencies, together with interpretative factors among 'street-level bureaucrats' and other stakeholders, still

---

[9] The 'Mediterranean syndrome' is characterised by structural deficiencies common to most Mediterranean countries, such as corruption, the lack of comprehensive plans or programmes to combat environmental problems and poor co-operation between the various administrative sectors that hold competence in issues such as desertification (La Spina and Sciortino, 1993; Pridham, 2002).

continue to impose productivist or otherwise environmentally harmful interpretations of policy goals and practices. This occurs not only in relation to environmental policies but also to the management of natural resources more widely. Therefore, although decision-making at the operative level of land management - for example at farm-level - is crucial in determining the influence of policies on desertification, it is evident that the interpretative and administrative frameworks of policy implementation also need further exploration in order to explain the often unexpected impact of policies on desertification.

Assuming that the process of implementing a policy carries a decisive role in defining policy outcome, whether the way these policies affect and shape land management related decision-making is either beneficial or harmful in terms of desertification remains uncertain in the lack of more specific contextual explorations. This book attempts to fill a gap in understanding the full challenge and particular obstacles for implementation of policy initiatives (like the NAPs) left by research conducted to date. Not only is there a distinct lack of attempts at identifying and understanding the links between existing individual policies and desertification at a Southern European scale, but the socio-political contexts of policy implementation and the social interactions and networks governing policy implementation and landuse at national, regional and local level also remain little understood.

The following chapters will, therefore, address the various policy processes that lead to both *expected* and *unexpected* impacts in more detail, not only identifying a range of relevant policies, but also the relevant stakeholders, their interconnections and competing interests and interpretations guiding policy implementation and natural resource management. Desertification narratives and trajectories of the four case studies analysed in Part II, and the interpretative analysis of actor roles, power and interests in Part III, are, therefore, attempts to provide a more informed and informative analysis for desertification mitigation in Southern Europe that will suggest meta-level concepts that may help better understand the policy dimension of desertification.

4. Desertification and policies in Spain: from land abandonment to intensive irrigated areas

J.J. Oñate and B. Peco

The aim of this chapter is to analyse interlinkages between policies and desertification in Spain, with a specific focus on the Guadalentín valley in the region of Murcia. The chapter is structured into six sections. A brief introduction is given in Section 4.1, while Section 4.2 discusses methodological approaches used in our study, including discussion of the rationale behind our approach and describing the research tools used. The historical and institutional framework in which the researched reality is embedded is then analysed in Section 4.3. Section 4.4 describes the main characteristics of farms surveyed in the area and assesses the not always similar perceptions of desertification between different stakeholder groups and farmers. Section 4.5 compares and integrates results from stakeholder and farmer surveys conducted in the area with regard to policy-related landuse changes, processes and effects experienced in the area over the last 20 years, while the main conclusions of the research are outlined in Section 4.6.

4.1 Introduction

As discussed in Chapter 2, the issue of desertification has been on the Spanish political agenda since the very beginning of international efforts to address the problem. Despite this, however, Spain is the only Southern European country that has not submitted its NAP to the UNCCD (as of December 2004). By examining the case of the Guadalentín valley in the region of Murcia (see Figure 4.1 below), this chapter will analyse this paradoxical situation with a specific emphasis on the often contradictory goals of economic development and environmental conservation.

As early as 1981, and following the guidelines of UNCOD (see Chapter 2), Spain launched the *Project to Fight Desertification in the Mediterranean* (Pérez and Barrientos, 1986), which, consistent with the interpretation of 'desertification' at the time, restricted its spatial focus to semi-arid south-eastern regions of the country and its thematic scope to the problem of soil erosion alone. In spite of the evident shift of emphasis in the new definition of desertification adopted by UNEP after 1991 (see Chapter 2), the Spanish approach to desertification did not change in respect to three fundamental issues: first, misleading analysis of causes and processes behind desertification and of their relative importance through time, leading to an unbalanced response in terms of contemporary actions against the main drivers of the problem; second, under-valuation of arid and semi-arid environments (largely comprising traditional regimes of dryland agriculture of low economic profitability) based on the arbitrary conceptualisation of these areas as 'degraded' environments; and, third, subordination of desertification mitigation measures to water management as the main instrument for the social, economic and spatial development of the country, actively promoted and financed by public intervention and supported by the Spanish water policy lobby that has remained anchored in its traditional concepts and approaches.

All these three issues are prominent in the 3,300 km² Guadalentín River basin, situated mostly in the south-eastern Spanish region of Murcia (Figure 4.1) which forms a paradigmatic case for the above-mentioned paradox. The Guadalentín has been one of the target areas for long-term research on physical processes involved in desertification conducted under the EU-funded MEDALUS projects (see Geeson *et al.*, 2002). In this region, efforts have also been made to understand the socio-economic factors behind desertification in terms of landuse and management changes (Barberá *et al.*, 1997; López-Bermúdez *et al.*, 2002; Romero-Díaz *et al.*, 2002) and, in a broader context, policy implications of these changes (CEC, 1997). Despite all this existing research in the area, a complete understanding of why many national and EU policies have failed to rectify desertification problems, let alone why some policies have even exacerbated desertification processes, is still lacking.

The Guadalentín represents one of the most severe cases of desertification in Southern Europe, with several issues at the centre of the problem. First, surface and groundwater over-exploitation (CHS, 2001), soil salinisation (Pérez-Sirvent *et al.*, 2003) and natural habitat destruction (Martínez and Esteve, 2000), together with a large increase of irrigation agriculture in the valley in recent decades to currently 48,000 ha (CEH, several years), have

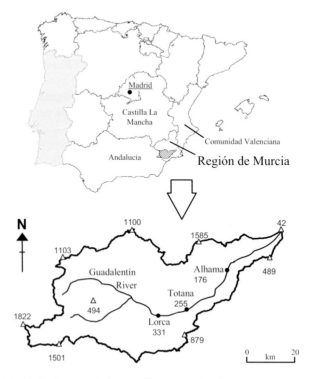

Figure 4.1: The Guadalentín basin case study area (Source: authors).

enhanced desertification problems. Second, the area suffers from intense erosive dynamics in the hilly dryland zones, rooted in historical landuses and present management changes acting on a sensitive combination of semi-arid climate and vulnerable soils (López-Bermúdez *et al.*, 1998). The dichotomy of dryland versus irrigation farming is anchored in a long-standing perception of the role of water and irrigation as drivers of rural development (Saurí and Del Moral, 2001), for which not only individual stakeholders are responsible but also institutional responses and management objectives. Design and implementation of, and response to, landuse-related policies in the area can be expected to be conditioned at all levels by long-lasting perceptions, to the extent that any analysis of their impacts on desertification must take into account the particular historical and institutional settings of the area.

In our opinion, understanding the context of how individual farmers' decisions on landuse and management have been (and continue to be) made, is key for the development of feasible strategies and policy options to fight desertification in the area. Therefore, our attempt under the MEDACTION project was to explore and characterise this context in order to adequately understand perceptions, decision-making processes, and policy responses by actors on the ground, thus focusing on society-driven aspects of the problem. Although this approach is not new (see, for example, Van der Leeuw, 1998; Lemon, 1999), the peculiarities of the Guadalentín area make it an appropriate area to investigate the socio-political dimension of desertification, since the complex and varied current desertification problems of the area cannot be fully understood or managed without consideration of the social context.

4.2 Methods

4.2.1 Rationale of the study

Our point of departure is that desertification problems in the Guadalentín are rooted in certain physical characteristics - notably a semi-arid climate, available groundwater and highly erodible metamorphic and sedimentary rocks - which historical trends of landuse and social and technological change have exacerbated. As Figure 4.2 shows, development paths before and after EU accession in 1986 have reinforced two related landuse change dynamics that have been key causes of desertification: expansion of irrigated agriculture in the lowlands and substantial changes to farming practices in the surrounding dryland areas.

Expansion of irrigated agriculture is a main driver for semi-natural habitat destruction and aquifer depletion in semi-arid climates, leading potentially to the desiccation of boreholes and aquifer salinisation, similar to processes described by Lemon *et al.* (1994) for the Greek Argolid Valley. The latter include both intensification and abandonment of agricultural practices as well as sudden changes in crop choices following more rewarding EU subsidies. Effects of such changes on erosion rates have also been reported elsewhere in the Mediterranean (Kosmas *et al.*, 1997).

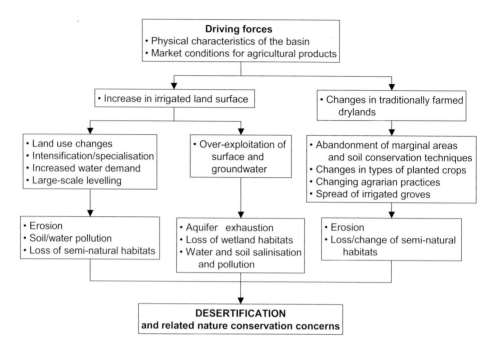

Figure 4.2: Historical trends in landuse, associated processes and desertification-related impacts in the Guadalentín Area (Source: authors).

On the basis of documented information and our own work, our first interest was to reconstruct the role of past policies in these processes, in particular national water and agricultural policies and, after 1986, also agricultural and structural policies implemented at EU level. Moreover, the role of a set of recent instruments that could have alleviated the problem was also investigated. Instruments applicable on private land, such as agri-environmental and agro-forestry schemes, were considered, together with broader policy instruments such as landuse planning policies and the Spanish NAP[10].

4.2.2 Methods of inquiry

The information basis for our research consisted of 25 in-depth interviews with relevant stakeholders on the issue of desertification and 104 field questionnaires delivered to a randomised sample of farmers in the area. An extensive literature review of policy work already conducted in the Guadalentín Area, as well as an analysis of the range of past and present policies that should form the focus of the investigation, were carried out in advance (Cummings *et al.*, 2001a, 2001b; Oñate *et al.*, 2001). Although at the scale of the entire valley the characteristic dynamics of different landuses are related and mutually reinforcing, the policies directly behind individual landuse changes are distinct. Irrigation has been used

---

[10] The role of instruments applicable on public land, such as hydrological corrective measures and forestry measures, were also considered, but they are not discussed here as they are not implemented via the farmers (see Oñate and Peco, 2005, for further details).

traditionally and continues to be the exclusive focus of a complex mixture of water and agricultural policies, mostly structural policy instruments co-financed by European funds and complemented by national and regional initiatives. Their main objective has been to address the water deficit in the area and to improve irrigation efficiency. On the other hand, market policy instruments of EU origin have targeted dryland agriculture since 1986, with objectives linked to maintenance of farming in a marginalised environment. The great differences in economic turnover and profit between irrigation-based and dryland farming are also reflected in differences with regard to successful policy implementation (Table 4.1; see also Chapter 2).

With regard to expert opinions, face-to-face semi-structured and taped interviews were conducted with selected governmental representatives at the national (6 interviewees) and regional (7) levels, farmers' organisations (6), academics (4), private corporations (1) and NGOs (1). This provided a variety of perspectives considered essential in obtaining a holistic view of the desertification issue. Existing contacts at the governmental level acted as starting points for the building of the interviewee network and facilitated access to most administrative officials. Using a snowballing methodology, some of these interviewees, in

Table 4.1: Past policy instruments linked to desertification considered in the Spanish case study area (Source: authors).

Irrigation farming
Water policies
- Law 52/1980 on the regulation of the economic regime of the Tajo-Segura water transfer
- 1985 Water Law
- R.D. 3/1986 on urgent measures for the hydrological planning of the Segura river basin
- 1999 new Water Law

Agricultural policies
- Structural measures (e.g. Reg. EEC/797/85, EC/2328/91, EC/950/97 and EC/1257/99)
- Off-farm investments for water supply and distribution (implemented through *Comunidades de regantes*)
- On-farm investments to enhance irrigation efficiency (farm modernisation aids)
- Law 9/1996 on urgent measures to ameliorate the consequences of drought
- Reg. EC/2078/92 on agri-environmental measures (incl. integrated control in irrigated vineyards)

Dryland farming
Agricultural policies
- CAP direct payments for cereals, olives and almonds (e.g. EEC/1035/72, EEC/2727/75, EEC/466/82, EEC/797/85 and EC/2000/96)
- Reg. EEC/1094/88 on set-aside of arable land
- Reg. EC/2078/92 on agri-environmental measures (incl. integrated farming of almond plantations and organic farming)
- Reg. EC/2080/92 on afforestation of agricultural land
- Reg. EEC/2159/89 on structural measures for the enhancement of dry fruit plantations

turn, suggested other contacts from farmers' organisations and private corporations (cf. Lemon, 1999). Data collection was conducted during 2002 (stakeholder interviews) and 2003 (questionnaire survey).

With regard to the farm questionnaire survey, its general design reflected the distinction between irrigation and dryland agriculture in the area. As a result, the sampling methodology followed a preliminary stratification into irrigation and dryland areas. Greater weight was assigned to the latter (77 dryland farmers sampled) than to the former (27 irrigation farmers), broadly reflecting the relative importance of each type of farming in geographical terms. Both sub-samples were further stratified according to farm size to better represent existing farm structures in the area. Thus, 13 questionnaires were delivered to irrigation farms smaller than 5 ha, while 14 questionnaires were delivered to irrigation farms larger than 5 ha. Similarly, 41 questionnaires were delivered to dryland farms with less than 20 ha, 25 to those between 20 and 50 ha, and 11 to farms larger than 50 ha. The general approach to identify policy effects on farmers' decision-making followed a policy-response model and was based on asking farmers, first, whether they were taking advantage of each specific policy instrument on offer and, second, what changes in landuse, crops or management techniques these policies had induced.

4.3 Historical and institutional framework

Information on historical trends affecting the region and the existing decentralised institutional framework derived from stakeholder discourses formed the context for an understanding of the dynamics of land management in the Guadalentín, setting the scene for our analysis of the effects of past policies on desertification in the area.

López-Bermúdez *et al.* (2002) reviewed the literature on the origins and development of interactions between drylands and irrigation agriculture during the 18th and 19th centuries. In that period, several waves of dryland agricultural expansion into hill areas of marginal quality resulted in serious problems of soil erosion in the uplands and flood damage in the lowlands. Irrigation agriculture, which dates back to the Bronze Age along the lower river courses in the area, remained limited in extent by technological factors governing water availability, in spite of repeated efforts towards expansion, including planned construction of several dams that were never built.

The beginning of the 20th century saw the issue of irrigation entering the political arena, with the consideration of water resources as key to social, economic and spatial development, and the Spanish state paying for the costs of the necessary dams and related infrastructure[11]. In a first development phase, technological advances secured access to surface water resources, which were exploited to the maximum for irrigation expansion. In a second phase, the arrival of submersible pumps enabled the exploitation of groundwater and further irrigation expansion. As a result, new opportunities for individual farmers and

---

[11] The lack of consideration for environmental dimensions of water management is one of the key axes of the so-called 'water paradigm' (Saurí and Del Moral, 2001), which has driven Spanish water policy since the beginning of the 20th century.

job-creating agro-business companies emerged, resulting in the fact that almost the entire southeast of the area was allocated to intensive horticulture and fruit production in development plans of the 1960s, leading to even greater increases in water demand for irrigation (Harrison, 1993). Off-site implications of the new model became evident with the construction of the Tajo-Segura transfer channel in the mid 1970s, which represents the third phase of technological development. The arrival of water from the centre of Spain contributed to further expansion of irrigated areas, as did expectations for further increased water availability in the future. However, an intense drought in the late 1970s revealed that the design parameters for the channel had set unrealistic expectations of water surplus in the donor basin (López-Bermúdez et al., 2002). As a consequence, Murcia never received the expected yearly 900 Hm^3 of water, with resulting increases in water deficit for the area although each new input of water is, theoretically, meant to be allocated in such a way as to avoid water deficits in specific areas (CHS, 2001)[12].

The expansion of dryland agriculture, meanwhile, triggered other types of problems. It probably caused the highest erosion rates in the 1940s, when cereal crops were promoted at the expense of existing forest areas (Barberá et al., 1997). The impact of technological development was again crucial, for example during the 1960s when the production of esparto (*Stipa tenacissima*) - a native grass species used traditionally in wickerwork - became unproductive due to competition from plastic and was substituted by almonds, carobs and figs. Intensive ploughing of soils of marginal quality along slopes was also facilitated when tractors became affordable. Although the 1950s saw the implementation of preventive afforestation measures in the uplands in response to severe flooding in lowland settlements, aggressive forestry methods, including terracing with heavy machinery and widespread planting of conifers, partly exacerbated the problems of erosion (Chaparro and Esteve, 1995).

In the late 1970s, droughts pushed dryland agriculture even more into marginality (López-Bermúdez et al., 2002). This trend became irreversible since new market opportunities for irrigation products arose thanks to the preferential agreement between Spain and the European Economic Community (EEC) in 1970. Access to, and demand from, the international market was opened completely after Spain's accession to the EEC in 1986 and its incorporation into the Single Market in 1992 (Pérez Yruela, 1995)[13]. This further increased government support for irrigation at the expense of drylands. The consequent abandonment of traditional soil conservation techniques accelerated the ongoing erosion problems (Cerdá, 1997), although in some areas a slow but constant process of shrub vegetation recovery has also been reported (Obando, 2002). From the early 1990s, only CAP direct subsidies have supported Spanish dryland agriculture with, at times, further dramatic effects for erosion (see below).

---

[12] The deficit is now officially estimated at 460 Hm^3, although according to one national administration official "considering illegal overexploitation it may reach 800 Hm^3".

[13] The volume of vegetable exports doubled between 1986 and 1996, representing in the mid 1990s nearly half of the entire Murcian foreign trade (INE, 2000a). Germany, the United Kingdom, France and the Netherlands are the key destinations for Spain's horticultural exports (84% altogether) (CESRM, 1997).

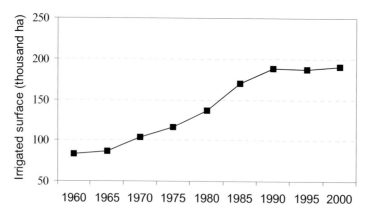

Figure 4.3: Regional trends in irrigated land in the region of Murcia 1960-2000 (Source: CEH, several years).

Overall, in less than two generations farmers have returned from forced emigration to industrial centres of Barcelona and Madrid, to live in the region with one of the highest growth rates of agricultural productivity in the country (CESRM, 1997). The spread of irrigated land has been part of a regional trend (Figure 4.3), now almost covering 31% of the Utilised Agricultural Area (UAA) in the region, more than twofold the national level (MAPA, annual reports).

The development of horticultural production and related activities has led to remarkable economic development in the area[14], diverting resources and attention away from economically increasingly marginalised dryland farms. Improvements in standards of living have led to a social momentum in favour of irrigation at the expense of dryland farming - a process described by several interviewees with the Latin expression *'animus regandi'*. Further, irrigation has also recently led to new and surprising associations in peoples' perceptions based on the assumed equations of water = irrigation and drought = desertification. Indeed, interviewees often argued that "irrigation is a constraint on the advance of the desert" or "the desert begins where lettuces end".

In addition to these historical developments, the institutional framework emerges as another important factor for understanding the present situation with regard to desertification in the area. Under the Franco dictatorship, both the Ministry of Agriculture (promoting landuse transformation) and the *Confederación Hidrográfica del Segura* (CHS, the River Board in the 18,815 km² Segura basin which regulates and supplies water resources), contributed to encouraging irrigation expansion in the area. Therefore, the 'irrigation problem' already existed well before the 1978 Spanish Constitution decentralised government, leading to a sharing of power between the central (hereafter 'national') and the regional governments or

---

[14] In the 1975-1996 period, the primary sector in the region grew at a rate of 4.3 %, while the growth of the entire regional economy was only 3 % (MINHAC, 2000). As a result, in 1999 the share of the primary sector in the regional Gross Added Value was 8.9 %, double the national figure of 4.2 % (INE, 2000a).

Comunidades Autónomas (hereafter 'regions'). As a result, since 1982 the Region of Murcia has become responsible for the legislative development and implementation of several issues of interest with regard to desertification (Table 4.2).

Table 4.2: Desertification-related issues covered by the Murcia Regional Government and year of transfer of responsibilities from the National Administration (Source: authors).

Year	Issues covered
1982	• Promotion of regional economic development under national policy objectives
	• Agriculture and rural development
1983	• Land planning and public works of Regional interest
1984	• Management of the environment
1985	• Design, construction and exploitation of hydraulic infrastructure
	• Water supply and sewerage
	• Management of nature conservation

In agriculture, the national role is now limited to leasing between EU regulations and regional performance and coordinating inter-regional initiatives, with real power lying, therefore, with the Regional Agriculture Department. The national level does, however, still have major power over water resources through the River Boards which, under the 1985 *Water Act*, must design their respective hydrological plans, administrate and control public water resources and uses, and maintain the infrastructure financed by them[15].

The key issue regarding irrigation and desertification issues (i.e. salinisation and water mismanagement) is undoubtedly the fact that the economic significance acquired by irrigation farming has given this type of farming unanimous support from the entire institutional and political spectrum. Although recognising the negative effects of the uncontrolled increase of irrigation on issues of institutional competence, even the regional departments of environment and physical planning act as subsidiary to the objective of irrigation expansion, claiming lack of effective powers to tackle the problem. This means that the current structural water deficit in the area has failed to become a political issue, with all actors now focused on expectations raised by the recently enacted National Hydrological Plan (NHP). This plan foresees a large investment in infrastructure for water transfer from the Ebro River basin (northeast Spain) to Murcia and neighbouring regions (MIMAM, 2001a), albeit in the face of heated debates and protests (Saurí and Del Moral, 2001). Expectations of new water supplies have spurred most of Murcian society to support the NHP, to the point where critics, such as environmental NGOs and some academics, have been branded as 'traitors' to the regional interests. Also hotly debated is the National Irrigation Plan, passed in April 2002 and planned to come into effect in 2008, especially as it foresees *no new*

---

[15] Having formerly belonged to the Ministry of Public Works, authority over water resources was transferred to the newly created Ministry of Environment in 1996, a change with no visible effects on the environmentally-harmful approach to water management in Spain.

irrigation expansions for Murcia but just improvements of existing irrigated areas (MAPA, 2002). This would mean that water transferred from the Ebro should, in theory, only cover current water deficits in the basin and not be used for creation of new irrigated areas. Finally, since coastal tourism and recreation have emerged as alternative development options (MINHAC, 2000), many stakeholders in the region (including most interviewed farmers) are becoming increasingly suspicious of the 'real' destination of transferred water resources.

4.4 Farm types and perceptions of desertification

As this section will show, the above discussion has important repercussions as to how desertification in the case study area is perceived and defined by our interviewed stakeholders. The analysis in this section will focus on perceptions of desertification by both farm and non-farm stakeholders and reveals, first, how farmers tend to perceive the issue of desertification in a more 'simplistic' way than non-farm stakeholders and, second, how farmers' perceptions of desertification are linked mainly to water and productivity issues. Before analysing stakeholder perceptions of desertification in Section 4.4.2, the next section will briefly outline the characteristics of the two main farming types that exist in the case study area as a basis for contextualising the different desertification perceptions with regard to the two key types of farming in the area.

4.4.1 Characteristics of surveyed farms

Two distinct types of farmers can be recognised in the Guadalentín target area, namely irrigation farming and dryland agriculture. This is not to say that mixed holdings are non-existent, but rather that most farms lie within one of these two categories. Their characteristics are important for discussing the changes that both have experienced in the past, for understanding different responses and perceptions to desertification, and for giving a basis for understanding future prospects.

Differences between both farming types are particularly pronounced with regard to farm area, number of fields, and property size, which are all larger in dryland farming areas. In irrigated areas, meanwhile, indicators for tenanted farmland, distance between parcels, income dependency on the farm, labour force and contracting of external services are all larger (Table 4.3). Significantly, both types of farmers work with farmers' unions and cooperatives to an equally high degree. They also share low levels of formal education, with a minority holding secondary degrees (12% [dryland farmers] and 4% [irrigation farmers]), although many have followed training courses, particularly irrigation farmers (71%).

The range of different landuses among surveyed farms (Figure 4.4) reflects the general importance of permanent crops found as the only crops on 80% of surveyed dryland farms, while 20% of holdings are mixed farms with some arable crops. Almonds, vineyards and olive plantations predominate over cereals, and in the drylands there are no purely arable farms. On irrigated farms, arable crops dominate (mainly horticulture) although these are never grown exclusively, and most holdings can be classified as mixed farms (52%), followed by those exclusively dedicated to permanent crops (37%). Vines are the most frequent irrigated

Table 4.3: Structural and household differences between surveyed dryland and irrigation farms in the Guadalentín Target Area (Source: Spanish farm survey).

	Dryland farms	Irrigated farms	Statistical test used	p
Number of farmers	77	27	-	
Sampled area (ha)	2509.21	276.55	-	
Total area of property (ha)	88.9 %	61.0 %	-	
Total rented area (ha)	11.1 %	29.0 %	-	
Average farm area (ha)	32.60 ± 37.4	10.20 ± 14.3	U=335.0	0.0000
Average number of fields	5.37 ± 5.6	4.70 ± 4.1	n.s.	n.s
Average distance between fields (km)	3.35 ± 5.5	4.66 ± 5.5	U=684.0	0.0350
Full farming income dependency (farmers)	52.0 %	92.3 %	χ^2=12.4093	0.0004
Labour force (average WFU)	1.28 ± 0.7	2.43 ± 2.17	U=557.0	0.0000
Union membership (farmers)	66.2 %	66.6 %	n.s.	n.s.
Cooperative membership (farmers)	80.0 %	88.9 %	n.s.	n.s.
Contracting of external services (farmers)	36.4 %	74.1 %	χ^2=9.9718	0.0016

χ^2 : Chi square test; U: Mann-Whitney U test; n.s: non-significant

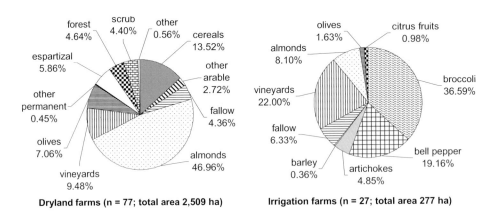

Dryland farms (n = 77; total area 2,509 ha)   Irrigation farms (n = 27; total area 277 ha)

Figure 4.4: Share of different landuses among surveyed farms (Source: authors).

permanent crop, followed by almonds and olives. Linked to the problem of land abandonment in many economically marginal farming areas in the drylands of Spain, non-cultivated areas are common in the dryland farming area, with about 20% of surveyed land covered by *espartizal*, forest, scrub and fallow. Livestock rearing on semi-natural rough grazing was unimportant in our dryland farm sample, with just 120 goat livestock units (LUs) and 59 sheep.

Regardless of the type of farm, most respondents (dryland farmers 95%; irrigation farmers: 96%) declared their intention to continue with farming in the next five years. In both cases, more than a third declared their intention to have a successor to take up the business, suggesting that in the medium term not much change is to be expected in farming districts of the case study area.

4.4.2 Stakeholder perceptions of desertification

Based on the pronounced differences in farm types and farm characteristics outlined above, irrigation and dryland farmers hold distinctive views on desertification. All irrigation farmers, for example, pointed to 'lack of water and loss of productivity' as the main drivers of desertification in their areas, and this perception was also shared by 60% of the 77 surveyed dryland farmers. In particular, five out of six interviewed members of the Professional Farmers' Organisation (PFO) held this view. This linkage between aridity and desertification was, however, not exclusive to grassroots actors, as three interviewed national administrators (working in water management and irrigation planning), and four interviewees from the *Regional Agriculture Department*, also agreed with this perception. In our opinion, this perceived linkage between lack of water and desertification is the basis of the above-mentioned 'cultural' argument supporting irrigation expansion in the area (e.g. "people irrigate because it is part of their culture" in the words of one interviewee), as well as the main support of the 'ethical' argument for further water demands for irrigation ("irrigation is a restraint on the advance of the desert" as one PFO member declared). Lending further support to this relatively unanimous perception of desertification is the fact that more than half of the 27 surveyed irrigation farmers acknowledged salinisation problems in their fields, caused by groundwater extraction from excessive depths.

'Unsustainable use of water resources' was the next most commonly mentioned conceptualisation of desertification by dryland farmers (13%). This coincides with perceptions expressed by two interviewees falling into the 'environmentalist' camp (one green group representative and a researcher in natural sciences). Significantly, water is also at the heart of this perception, but the emphasis here is on over-exploitation by irrigated agriculture as perceived by a significant proportion of dryland farmers who, in comparison, see their own farming methods as more benign with regard to exacerbating desertification processes. Further, eight percent of dryland farmers also agreed with five interviewed non-farm stakeholders that desertification can be understood as 'humans *deserting* the area after falling land productivity and agricultural abandonment' - in other words, equating 'desertification' with land abandonment and not with physical land degradation processes[16]. Non-farm stakeholders holding this view argued that this problem occurred mainly in dryland areas, driven by aridity and worsening market conditions, but this view was not shared by all dryland farmers. Thus, while 'loss of soil productivity due to human causes' (mainly deforestation and inadequate agricultural practices) was seen as a key definition of desertification by four non-farm stakeholders, only four percent of dryland farmers agreed with this notion. Erosion was, however, mentioned as a problem by nearly 90% of irrigation farmers and by 64% of dryland farmers.

---

[16] This is similar to the French term 'desertification' which means 'land abandonment' (Buller et al., 2000).

Finally, although desertification was understood as a 'global process of environmental degradation' by five non-farm stakeholders, this was only supported by two percent of dryland farmers. This highlights that grassroots actors in Spain may tend to be more 'removed' from abstract and global conceptualisations of the desertification issue than non-farm stakeholders. Inevitably, this will have repercussions for how policies to combat desertification are perceived (and implemented) by farm and non-farm stakeholder groups, an issue we turn to in the next section.

4.5 Policy effects over the past 20 years

On the basis of interviews with selected stakeholders and the farm questionnaire survey (see above), this section analyses the effects of implemented policies over the last 20 years on farmer's decision-making regarding landuse and management on their holdings, with potential repercussions for alleviating or exacerbating existing desertification problems. Table 4.4 highlights the different policy instruments that have targeted dryland and irrigation farming in the case study area. In the following, policy instruments focusing on irrigation and dryland farming will be discussed separately (Sections 4.5.1 and 4.5.2), while landuse planning policy issues and the Spanish NAP are analysed in Section 4.5.3.

Table 4.4: Implemented policies targeting dryland and irrigation farming and surveyed farmers affected by these policies (Source: Spanish farm survey).

Dryland farming (N=77)	**Beneficiaries**
CAP direct payments	
• Olives	9
• Cereals	8
CAP subsidies for dry fruit plantations	31
Agri-environmental measures	
• Almonds (integrated farming)	19
• Almonds (organic farming)	13
Agro-forestry measures	1
Irrigation farming (N=27)	
Water policies	
• Water concessions in 1980	12
• Water concessions after 1980	8
• 1985 *Water Act*: legalisation of groundwater extraction	9
• Groundwater concessions after RD 3/1986	4
Agricultural policies	
• Legislation to improve on-farm irrigation water efficiency	2
• Subsidies to address effects of 1996 drought (Law 9/1996)	11
• Agri-environmental measures for integrated farming in vineyards	17

4.5.1 Irrigation farming

All interviewed stakeholders pointed at the Tajo-Segura water transfer channel as the main development underlying irrigation expansion after 1980, the time period when most permits for running water were allocated. As we have seen, demand by the newly developed irrigation areas quickly exceeded water supplies, largely due to droughts in the 1970s. As a result, groundwater exploitation was provisionally authorised, and enhanced technology enabled wells to be drilled deeper in subsequent years with groundwater extraction becoming a permanent practice in the area.

Although the 1985 *Water Act* legalised most groundwater pumps and implemented a system of extraction permits, interviewees commented that this legislation could not prevent the spread of illegal groundwater extraction activities. As irrigation business grew during the late 1980s, an unregulated water market emerged with permit owners illegally selling water they were allocated but did not use. Several interviewees mentioned the lack of responsibility by the regulating body, the Water Board (CHS), in this initial stage of irrigation expansion. It should also be noted that the irrigation issue was further exacerbated in the 1980s by access for farmers to structural aid for farm modernisation after Spain's entry into the EEC. Although this helped increase farm productivity, it also indirectly promoted irrigation expansion with consequent increased pressure on water resources. This highlights that Spain's EU accession did not serve to limit or control irrigation expansion, especially as horticultural production was never covered by CAP direct payments.

The new 1999 *Water Act* prohibited the building of new wells but could not reverse the trend. As in the rest of Spain, rural policy design and implementation continued to be dominated by a 'productivist' ethos, in which agriculture and water policies combined to modernise and improve the competitiveness of irrigation (Wilson *et al.*, 1999; Sumpsi, 2001). As a result, overexploitation of groundwater resources intensified (López-Bermúdez *et al.*, 2002), impacting on soils due to salinisation, and on remnant wildlife habitats due to the drying up of natural wells and wetlands and through habitat destruction. During interviews, elusive answers were provided by regional officials on the question of these impacts and their tacit causes, and even the generally accepted fact of contemporary expansion of irrigated land was blatantly denied (see Oñate and Peco, 2005). This latter point has been a contentious issue throughout our research. Farm interviewees and personal observation clearly suggested that irrigated areas had substantially increased over the past 20 years (see below), and, similarly, Barberá *et al.* (1997), in their analysis of landuse changes in the Guadalentín area during 1950-1996, also confirmed a substantial increase of irrigated crops from 9% of the entire surface to 19% during that time period. Surprisingly for us, however, statistics from the Regional Government appeared to support statements from regional officials and did not confirm increases in irrigated land surface. Based on these statistics, and as Figure 4.5 illustrates, cropped surface in the five municipalities included in the Guadalentín area (Aledo, Alhama, Librilla, Lorca, Puerto Lumbreras and Totana) is portrayed as having remained almost constant since 1989, both for irrigated and dryland crops. As highlighted in Section 4.5.3, these discrepancies between fact and fiction make it very difficult for any decision-maker to tackle the issue of desertification in the Guadalentín area through effective policies.

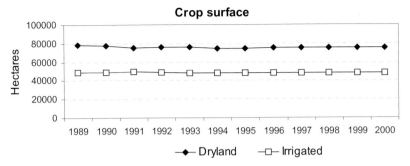

Figure 4.5: Cropped surface (arable and permanent crops) in the Guadalentín area 1989-2000 (Source: CEH, several years).

Another frequently mentioned symptom of water overexploitation was the lowering of river flow rates, weakening the system's capacity to absorb sewage from the growing population[17]. As a result, a leading Spanish newspaper recently commented that the Segura River was 'a sewer' (El País, 1999). In spite of being responsible for sewage control and water treatment, the regional government has only recently started to tackle the problem, and as one interviewed regional official put it, "works cannot be finished overnight". This highlights a certain lack of commitment by the authorities to solve the regions' growing environmental problems. Further, in addition to irrigation expansion, specialisation and intensification of irrigated agriculture have also taken place in the area[18], with associated land degradation effects such as diffuse pollution. However, provisions under the Nitrate Directive (676/91/EEC) have not been implemented (Izcara Palacios, 1998; see also Buller et al., 2000).

Beyond irrigation issues, the survey also investigated the opinion of stakeholders with regard to effects of agri-environmental and agro-forestry schemes, especially as these policies could play a crucial role for combating desertification (see Chapter 2). The first agri-environmental programme in Murcia (1994-1999) covered nearly 1,550 farmers (2.7 % of total), 40,000 ha (9 % of arable area), and cost about € 9 million. Unfortunately, a breakdown of figures for the Guadalentín area is not available. As in many areas of Spain (Peco et al., 2000), interviewees from the *Regional Agriculture Department* blamed the limited implementation of this scheme on budget restrictions, "especially since in Murcia irrigation is a top priority" as one of the interviewees argued. Passive resistance against agri-environmental schemes was also related to the conflict with predominant 'productivist' orientations and to the high transaction costs of AEPs (see Wilson, 2001, 2002). Although agri-environmental initiatives were considered useful for avoiding erosion and reducing agro-chemical usage, many

---

[17] According to the 2001 census, regional population has increased by a quarter since 1981, more than threefold the national figure of 8% (INE, 2002).

[18] Intensification is reflected in the 1993 figures for the ratio between Standard Gross Margin and UAA, which was 944 €/ha, and for that between Gross Margin and Annual Labour Unit (ALU), which was over 11,000 €/ALU, twofold and threefold respectively the national figures (CESRM, 1997). Indeed, in 1993 horticulture and fruit growing represented 62% of total agricultural production for the region, reflecting a much higher specialisation than in the rest of Spain (28%) (CESRM, 1997).

respondents claimed that the present CAP (including the recent Agenda 2000 reforms) does not adequately support farmers, thereby explaining delayed and low-level implementation of these policies.

The most successful agri-environmental measure has been integrated pest control on white grape vineyards, a measure targeted by most education and training efforts. However, the desertification mitigation role of this measure was considered limited beyond the positive effects of lower chemical usage, since no detailed requirements concerning 'good agricultural practice' were included in the specifications for this measure. Moreover, as one interviewee argued, the lack of a national regulation on integrated production was said to "facilitate the concealment of commercial production under an environmental façade", an opinion also shared by environmental groups.

The effects of past policies on irrigation farmland described by non-farm stakeholders were corroborated by the results from the farm survey, which showed that water policies related to the allocation of water concessions, as well as agricultural policies, led to increased irrigation use and efficiency. By 2002, up to 70% of the 27 surveyed irrigation farmers had increased their irrigated area since the time they installed irrigation, with an average increase of nearly 9 ha per farm (s.d. = 10.38), providing additional evidence that irrigation expansion has, indeed, taken place in the area (see above). Considering that the average area of the 27 farms is now just over 10 ha (see Table 4.3 above) this increase has been very dramatic.

The changes undertaken by both farmers who were already irrigating before 1980 and those who converted to irrigation post-1980, after water concessions from the Tajo-Segura transfer channel were granted, resulted in an overall expansion of irrigated area. The size of the original irrigation area was significantly higher for pre-1980 farmers (N = 19; 6.22 ha, s.d. = 7.41) than for post-1980 ones (N = 8; 2.31 ha, s.d. = 1.53; U = 32.0; p = 0.035). However, the latter were able to increase the original area to a greater extent than the former, up to the point that both groups of farmers had equivalent irrigation areas (in 2002) of around 10.8 ha on average (10.79 ha, s.d. = 15.85 and 10.82 ha, s.d. = 9.32, for pre- and post-1980 irrigation farmers respectively). It is interesting to note that not all 19 pre-1980 irrigation farmers were granted new water, and that among those 12 who were, the concession was considered 'insufficient' in 75% of cases. In spite of being illegal, existing wells (the common source of water before the channel was built) were deepened by 3 of these farmers and 4 of them admitted to illegally opening new wells in order to obtain water not legally allotted. Accompanying changes were the replacement of traditional flooding methods with drip technology (which was carried out on five of these farms), and the introduction of more profitable horticulture at the expense of bell pepper, cereals and cotton, together with the expansion of vineyards. The same story is true for the eight post-1980 farmers who began irrigation thanks to the new concessions linked to the Tajo-Segura transfer channel. These were considered sufficient by only 25% of these respondents, a circumstance that moved some to illegally open new wells (3 cases) or deepen existing ones (2). Significantly, drip technology was already present as an irrigation method in more than half of these post-1980 irrigation farms.

The impact of the 1985 *Water Act* on irrigation farms was low in the light of our survey. Less than a third of the 27 irrigation farmers legalised their wells, while the reasons why the majority did not do so remained vague (11% of the respondents did not answer the question). Farmers' reluctance to clearly indicate their water sources was anticipated in the stakeholder interviews, which emphasised the lack of effectiveness of water authorities in controlling the semi-legal water market emerging in the area (see Oñate and Peco, 2005).

After Spain's entry into the EEC, the joint effect of water and agricultural policies continued to encourage landuses that were potentially enhancing desertification in the area. New groundwater concessions in 1987 benefited nearly 15% of the surveyed farms, and at least one farmer admitted to having expanded his irrigated area as a consequence. Mirroring the experience of the 1985 *Water Act*, just 11% of farmers legalised their wells in 1987 (and 9 farmers introduced drip irrigation), the former a prerequirement for being granted new concessions. However, reluctance to answer questions on water sources was found in eight cases, suggesting that the amount of illegally obtained irrigation water is probably higher than the figures here suggest.

Many non-farm stakeholders argued that substitution of traditional irrigation methods (flooding) with drip irrigation has been policy driven, but evidence for this remains questionable on the basis of our farm survey data, since only 3 farmers acknowledged to have received public assistance for such change. Nevertheless, drought-amelioration assistance in 1996, received by almost 50% of surveyed farms, encouraged the introduction of drip irrigation on about half of the farms. Either directly or indirectly, policies aiming at off- and on-farm enhancement of irrigation efficiency have been, therefore, the most successful at reducing per-hectare water consumption for irrigation, with just less than a quarter of farmers still flood-irrigating their fields in 2002. In addition, nearly half of the farm respondents were irrigating exclusively by drip irrigation in 2002, while none had done so when they initially installed irrigation. Continued irrigation expansion has, therefore, co-existed with technological enhancement of irrigation efficiency, with 'saved water' having been re-employed by most farmers in expanding or intensifying their irrigated area.

As also pointed out by stakeholders, agri-environmental schemes targeting irrigation land seem to have had little impact on desertification, beyond the compulsory reduction in chemical input under integrated pest control measures. This was the most successful measure among surveyed farmers (13 farmers with 62.15 ha, 22.5 % of surveyed irrigated area), which was adopted mainly by farmers owning irrigated vineyards (10 farmers with nearly 30 ha). Yet, beneficiaries of these agri-environmental payments mentioned no changes in management techniques regarding water use. Although participation in integrated control measures for vineyards seems to be influenced by the amount of cropped surface, greater among participants (on average 3.47 ha; s.d.= 2.03) than among non-participants (2.45 ha; s.d.= 0.668; $U = 17.0$; $p = 0.039$), most of the participating farmers suggested that their main reasons to enter the measure was because it was recommended by the local cooperative, and because complying with this suggestion would put themselves in a better position vis-à-vis future CAP reforms. This crucial role of cooperatives in influencing farmers' decision-making is also clear regarding the size of irrigation expansion in the last 20 years, with cooperation members expanding to a greater extent than non-members (U = 7.500; *p* = 0.011) (see also Hagedorn, 2002).

Overall, few irrigation farmers linked their decisions regarding changes to landuse or management techniques to the policy environment. As stakeholders remarked, irrigation dynamics are the result of a complex mixture of water and agricultural policies favouring irrigation farming which operates in a favourable market for horticultural products. As a result, most surveyed farmers argued that the main reasons for landuse changes were related to market opportunities (especially for crop choices) and the need to save water (in the case of irrigation techniques). This shows that with regard to drivers of desertification in irrigated farm areas linked to salinisation and overexploitation of groundwater resources, increased profitability emerges as the key driving force for change and *not* the policy environment. At the same time, the policy environment has done little to help desertification mitigation. Nonetheless, over half of the 15 farmers who declared salinisation problems in their fields acknowledged that this problem was linked to the excessive depth from which they extract groundwater, suggesting some awareness of the underlying causes and issues surrounding desertification processes in irrigated areas. Vines were identified as the main affected crop, with an average surface of 4.4 ha/vineyard suffering from soil salinisation. Surprisingly, erosion was also mentioned as a problem by 93% of respondents, despite of irrigation farming being almost exclusively located on flat areas.

We have seen, therefore, that irrigation farmers' visions of desertification problems tend to be profit- rather than policy-oriented. Interview respondents from farmers' organisations argued that when surface water becomes scarce, and in order to afford investments made to introduce new irrigation, they "were obliged to extract groundwater", thereby exacerbating desertification. Respondents particularly complained about the lack of coordination between the *Regional Agriculture Department* and the CHS in managing the dynamics of irrigation expansion, and that large agri-business companies always seemed to be favoured by the fiscal and financial environment to the neglect of the small farmer. Farmers, therefore, saw irrigation as the only development possibility for them, and greater availability of surface water was unanimously seen as the solution to the problems in the area - a situation very different from that in the drylands of our study area, which is the focus of the next section.

4.5.2 Dryland farming

Two types of drivers, related more to *policies* than to economics, seem to have influenced desertification-related landuse and management changes in dryland farming areas in the Guadalentín. On the one hand, there are those policy instruments (mainly CAP-related) specifically targeting dryland crops, such as cereals, almonds and olives. On the other hand, the mostly hilly areas of the drylands have seen the *indirect* influence of policies behind irrigation (see above), both in terms of uncontrolled invasion of former dryland areas by irrigated citrus fruit plantations and through the lack of social recognition given to the drylands, as government attention is focused more on irrigation farming at the expense of dryland farming.

Most interviewed non-farm stakeholders agreed in dating the beginning of contemporary changes in the drylands back to the early 1980s, before EU accession. Prospects about forthcoming CAP direct payments encouraged expansion of cereal crops and almond plantations in previously abandoned areas, at the expense of recovering scrublands that had helped alleviate desertification processes. In this sense, creation of new fields in former scrubland areas, as well as abandonment of soil protection techniques on traditionally farmed land due to intensification, were mentioned by several interviewees as key drivers of erosion in the drylands. However, the precise impact of these developments was site-dependent, since a particular change may lead to effects of varying gravity depending on site conditions (slope, geology, etc.) and the continuation (or not) of soil protection techniques such as ploughing along contour lines, terracing and gully management. New almond plantations were particularly harmful for erosion, as they were often planted after substantial surface levelling with heavy machinery that destroyed less erosion-prone traditional terracing systems (see also van Wesemael et al., 2003).

Since EU accession in 1986, changes in crop patterns have continued, with economically marginal cereals further declining at the benefit of permanent crops including almonds, olives and vineyards. However, CAP set-aside subsidies after 1988 were mentioned as a beneficial policy in terms of desertification by some interviewees, as set-aside allowed the natural recovery of protective vegetation, thereby reducing erosion in affected fields. Yet, set-aside was also claimed to have been a driver for erosion on fields where farmers had still been undertaking soil protection measures. In this sense, more efforts in specific targeting of these subsidies would have been required.

The 1992 CAP reform accompanying measures, meanwhile, were seen by interviewed stakeholders as having largely failed to help combating desertification, because of the specific design of these measures and due to the long delays in implementation (Buller et al., 2000; see also Chapter 2). Among these measures, cereal extensification was mentioned as the most successful in the Guadalentín (with around 4,000 ha enrolled), while ploughing along contour lines was mentioned as the change in landuse practice with the most tangible effects with regard to erosion prevention. Integrated pest control and organic farming measures only saw limited uptake and, beyond the positive effects of lower chemical usage, interviewees argued that these policies only had a limited role in combating desertification, especially since no detailed requirements concerning 'good agricultural practices' were included in these measures (see also discussion on irrigated farming above).

The agro-forestry scheme had an even more limited budget allocation than the agri-environmental package. Yet, the perception among non-farm stakeholders was that agro-forestry measures could have been highly positive in terms of retiring low-productive intermittently cultivated drylands often prone to desertification, but only about 10,000 ha have been planted on former farmland in the entire region based on this measure (specific Guadalentín basin data unavailable). Further, the agro-forestry policy was also questioned by stakeholders because of design deficiencies, particularly its approach focusing on productivity (which is low in a semi-arid climate) and the lack of a landscape-based approach that could have prevented creation of a mosaic of unconnected forested patches with little overall environmental benefits.

The so called 'subsidy culture' was identified by most non-farm stakeholders as a potential threat concerning desertification, especially in light of suggestions for reduction or even complete abandonment of the subsidy regime by 2006-2008. As one interviewee put it, "farmers accustomed to subsidies might abandon soil conservation practices, which for better or worse, they are performing today". This was a vision unfortunately confirmed by one respondent from a farmers' organisation, who argued that "they [farmers] go into these schemes because it helps in monetary terms, not so much because of principles".

Nevertheless, environmentally-oriented stakeholders saw the extent of desertification problems caused by current dryland agriculture as much smaller than that caused by the highly visible and uncontrolled invasion of footslopes by irrigated permanent crops (see also Barberá *et al.*, 1997). Thanks to the increasing mobility of water based on technological developments (see above), irrigated citrus fruit plantations are at present expanding into footslope areas of the Guadalentín. Agro-business companies and individual farmers are the main agents behind this process. These actors, having bought low-priced dryland, invest in new water allocation measures to install irrigated plantations, thereby often eliminating erosion-mitigating structures such as terraces, with potentially devastating effects on desertification processes. Heavy machinery can easily work the highly mouldable metamorphic and sedimentary rocks to create a levelled surface that, despite the lack of soil, has enough fine particles to feed the crop. As one interviewee put it, "agriculture is no longer linked to soil which just acts as a physical base". After the water supply ends or the soils become too salty, the transformed plots are usually either re-converted into dryland or directly abandoned and left to erode.

Possibly the most far-reaching effects of irrigation expansion in the area, at least in human terms, seem to have been on social awareness and culture in the drylands. In this sense, human, technical and economic resources focusing on problems and demands of the drylands have been kept to a minimum in comparison to those targeting the more prosperous irrigated areas. As one agricultural official put it, "contact with dryland farmers is minimal, and is only made to monitor the received subsidies". Interviewees mentioned that the CAP and the EU were criticised for not being sensitive enough to the problems of Southern European countries, although the positive role of subsidies in maintaining an otherwise marginal activity was also mentioned. As a result, and in the opinion of non-farm stakeholders, locals have seen how the legacy of their ancestors has become undervalued in both economic terms and with regard to government attention. In the absence of such social recognition, farmers' management decisions tend to be more drastic, seeking the most rewarding subsidy regardless of good agricultural practice. As one interviewee argued, "from a desertification perspective, social change operates in a negative way for dryland farmers and their successors".

The farm survey data shows that dryland agriculture in the area is a part-time activity, mainly maintained by CAP subsidies and with structural and economic factors negatively influencing future prospects. In addition, erosion is a recognised problem linked to desertification in nearly 67% of the 77 surveyed farms, and defined as 'severe' or 'medium' by 40 farmers. Surveyed dryland farmers have been characterised as 'low farm-income units' for the last 20 years (see Table 4.3 above), with nearly 50% currently complementing agricultural income through other sources of revenue, mainly transport, industry, tourism and construction activities. Just 13% of dryland farmers have increased farm size over the past 20 years, reaching 30.2 ha on average at present and consisting mostly of freehold land. This farm size, clearly insufficient for economically viable cereal cropping in an adverse environment[19], has nevertheless facilitated the maintenance of permanent crops such as almonds, olives and vineyards, for which regional yields are higher than national averages[20]. No single surveyed farm grew cereal crops exclusively, but 24 farms exclusively grew almonds, three farms had only olives, two just vineyards, 34 had various combinations of almond and olive plantations and vineyards, and 14 farms combined permanent and arable crops.

CAP subsidies for these crops have played a crucial role in maintaining cultivation, although, paradoxically, less farmers than those who declared to grow them admitted receiving subsidies. Reception of CAP direct payments for cereals and olives was declared by only eight and nine farmers respectively, while 12 and 33 farmers respectively declared growing these crops. Similarly, receipt of subsidies for dry fruit plantations was declared by only 31 farmers, in spite of the number of farms growing them being 70. The apparent explanation for this paradox could be found in the role of the local cooperatives, which comprise about 80% of our sample and forming their main source of advice regarding available subsidies. Further, many farmers said they relied on their cooperative for almost all financial aspects of their farms and, therefore, were not fully aware of particular subsidies they had applied for. This shows that the role of the cooperatives went well beyond simple advice, and that, to some extent, farmers had ceded part of their decision-making powers to the cooperatives[21] (see also Chapter 7 on Greece for similar patterns and problems).

This has meant that the influence of CAP subsidies on crop choices (and thus on desertification) over the last 20 years, considered as 'substantial' by non-farm interviewees (see above), could not be verified at the grassroots level on the basis of our farm survey. However, the role of these subsidies in maintaining farming activity is clear in the light of poor crop yields and the fact that, without exception, farmers mentioned subsidies as the main reason behind their applications for different policy instruments[22]. Barberá et al. (1997), in their analysis of landuse changes in the Guadalentín area during 1950-1996, confirm the decrease of arable crops in the drylands (from 37% of the entire surface to 15%) and the concurrent increase of permanent dryland crops (mainly almonds, from 5% to nearly 11%). Unfortunately, available statistical data from the Regional Government on the evolution of crops and non-arable surface in the Guadalentín municipalities do not shed additional light on the issue (Figure 4.6). Thus, while these statistics show a marked decrease for dryland cereal crops in planted area between 1989 and 1999 (more than 14,500

---

[19] Average dryland cereal yields are less than 1000 kg/ha in the province of Murcia compared to the national average of nearly 3000 kg/ha (INE, 2000b).

[20] 273-350 kg/ha for almonds, 1016-1823 kg/ha for olives and 3898-4631 kg/ha for grapes (INE, 2000b).

[21] This also explains why most farmers did not mention changes in landuse or crops when asked about the impact of specific policy instruments. Nevertheless, a limitation of the inquiry method was also detected here: when asked, most farmers were unable to specify and date the crop changes they had made in the past, arguing that they did not remember the exact figures.

[22] Environmental considerations were only listed as first reason for subsidy application in the case of the agri-environmental measure for almond organic farming.

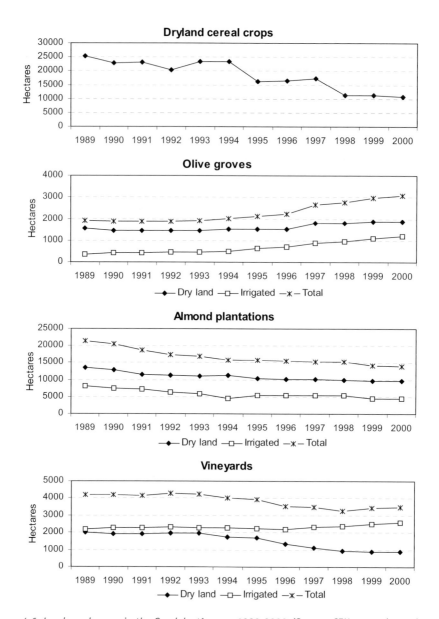

Figure 4.6: Landuse changes in the Guadalentín area 1989-2000 (Source: CEH, several years).

ha), and while an opposite trend can be observed with regard to olive groves in 1999 compared to 1989 (+ 1,150 ha; a trend particularly prominent since 1994-1995) which would seem to lend credence to our interviewees, these official statistics nonetheless suggest that almond plantations only occupied about 14,000 ha in 2000 (i.e. less than in 1989) and that this negative trend is present in both irrigated and non-irrigated plantations. Equally, according to these statistics, vineyards have lost around 700 ha since 1989, although the trend differs for non-irrigated vineyards which have decreased by 1,100 ha, and irrigated vineyards which have increased by nearly 400 ha (this divergence is particularly noticeable since 1993).

Despite of this contradicting information with regard to crop changes in the area, farmers' decision-making with regard to application for subsidies was influenced by several factors. Size of cropped surface, for example, played a crucial role for farmers applying for CAP subsidies for cereals, with farms which had applied having bigger areas of cereals (on average 43.36 ha, s.d.= 77.14; N = 8) than those who had not (5.08 ha, s.d.= 4.65, N = 3; U = 1.000; p = 0.024). Also, farmers applying for olive subsidies had larger areas of olive groves (on average 8.7 ha, s.d.= 7.80; N = 7) than those who did not (3.43 ha, s.d.= 3.65, N = 22; U = 34.0; p = 0.028). Participation in the integrated control measure for almond plantations, which included 19 farmers (total area 486 ha) appeared also to be influenced by cropped surface with participants having larger plantation areas (on average 25.57 ha; s.d.= 24.55) than non-participants (13.88 ha; s.d.= 10.71; U = 220.0; p = 0.017).

Farming-income dependency appeared to influence participation in the organic farming measure, which included 13 farmers with a total area 165 ha of almond plantations. Thus, farmers fully dependent on agricultural income were less frequent among participants than among non-participants (0% and 63% of farmers respectively; χ^2 = 13.08; p = 0.0001). Finally, the strong influence of cooperatives was again crucial for farmer participation in both almond plantation enhancement measures and integrated control measures. In both cases, members of the cooperatives were significantly more frequent among participants than among non-participants (94% among participants and 69% among non-participants in the case of the former, χ^2 = 4.954; p = 0.026; 100% among participants and 66% among non-participants in the latter, χ^2 = 6.589; p = 0.01).

Overall, the farm survey data emphasised that policies implemented over the last 20 years have mainly contributed to *maintain marginal farming activities*, which otherwise would have led to an accelerated process of land abandonment with potentially disastrous consequences for desertification in the area (especially through neglect of good farming practices). Direct payments for cereals and olives, as well as subsidies for the enhancement of almond plantations, have particularly played a crucial role, together with the more recent agri-environmental payments. Although not tested on the basis of our data, expectations of higher revenues from almonds and olives (both conventional and integrated/organic production) have most likely been a factor for expansion of the targeted crops at the expense of much less competitive cereals. This expansion of the 'subsidy culture', mentioned by stakeholders as a threat concerning desertification (see above), may have its worst effect in the context of the recent mid-term review of the CAP. As one academic interviewee put it, "farmers accustomed to subsidies might abandon soil conservation practices, which for good or bad, they are performing today". Thus, the effect of dryland farming-related policies on desertification, as anticipated from the stakeholder interviews, depends largely upon the concurrent development, and maintenance, of soil protection techniques for erosion prevention. Since none of the past policy instruments considered in this study entailed cross-compliance with regard to soil protection activities, it is not possible to assess the precise degree by which they might have prevented desertification in the past. Indeed, 15 of the

51 dryland farmers who acknowledged erosion problems on their farms declared that they did not carry out any erosion prevention measure, while 13 farmers mentioned ploughing along contours as their only practice to prevent the problem. Considering the economic marginality of most farms in the area, and the fact that the majority are part-time farmers, abandonment of erosion prevention practices, therefore, has to be seen as the potentially biggest threat for desertification in the drylands.

4.5.3 Landuse planning policies and the National Action Plan to Combat Desertification

In this last section, we wish to briefly come back to the wider policy arena and analyse stakeholder perceptions of landuse planning policies and the Spanish NAP. As mentioned in Chapter 2, Spain, as the last Southern European country, had still not delivered its NAP by the time of writing (January 2005), although the NAPs are a national requirement under the UNCCD framework (see Chapter 2). Only a working draft has so far been made available in 2001 (MIMAM, 2001b).

Significantly, only seven of the 25 interviewed non-farm stakeholders acknowledged familiarity with the draft NAP, eight had heard about it but were not aware of details, while the remaining ten were unaware of its existence. It seems, therefore, that there has been a pronounced lack of communication between those designing the Spanish NAP and wider society, resulting in the fact that many government officials are sceptical of the plan. In particular, respondents from the CHS and the Physical Planning Department were unaware of the NAP, as were all the surveyed farmers. The fact that the NAP is coordinated nationally by the Environment Ministry could explain the low interest among agriculture-related actors. Further, the temporal overlap of the NAP with the implementation of other national policies potentially *contributing* towards worsening desertification problems - in particular water and irrigation policies - was pointed out in the interviews as evidence of the lack of political will to tackle the root causes of the desertification problem.

Prospects for successful implementation of the Spanish NAP were seen as limited in light of past experience with landuse planning and related policies, which would have been "subsumed to the logic of irrigation expansion", as one interviewee put it. Paradoxically, expansion of intensive irrigation has been, in theory, prohibited since 1986, when the first regional landuse planning legislation was enacted, and subsequent hydrological planning documents have confirmed this. However, many respondents doubted whether these policies could have stopped or re-oriented irrigation expansion, given the existing powerful social and economic pressure surrounding the water issue (see above).

In Spain, landuse planning policies have traditionally focused exclusively on the territorial location of economic activity and necessary infrastructure arrangements (Oñate *et al.*, 2002a). This approach would partly explain why the total area of protected land in the region was reduced by nearly 11,000 ha in 2001, mainly around areas of agricultural and/or tourist interest (La Opinión, 2001). Legislation on nature conservation has not been able to prevent the visible invasion of protected areas by expanding irrigated plantations, a process denied, belittled and even supported by some respondents from the regional government. Although planning documents (MIMAM, 2001a; CHS, 2001; MAPA, 2002) mention the need to take environmental considerations into account when implementing irrigation projects, experience shows that this has never been a priority. In the absence of legislation prescribing environmental assessment at the strategic level of plans or programmes (Oñate *et al.* 2002b, 2002c), environmental impact assessment at the project level has neither been able to control irrigation expansion nor to alleviate its subsequent negative environmental effects.

4.6 Conclusions

Our research in the Guadalentín basin has illustrated clear impacts of past policies on land degradation. Overall, it could be said that desertification, as an environmental issue, has suffered from the so-called 'Mediterranean Syndrome' (e.g. La Spina and Sciortino, 1993), which makes the promotion of non-economic interests rare, or at least difficult, at the local level. Our results also show that the different dynamics of irrigated and dryland areas require a separate approach for anti-desertification policies.

Irrigation farming in the Guadalentín basin is a good example of highly productive semi-industrial agriculture, with small to medium capitalised and technologically well-equipped farms that are labour-intensive and highly dynamic without direct CAP payments. Spatially scattered field patterns per farm, high percentage of rented land and nearly full farming-income dependency by farmers all illustrate the structure of this sub-sector. Although few farmers related their landuse decisions to the particular policies mentioned during interviews, implemented water and agricultural policies have coexisted with a generalised increase of irrigated area and the introduction of dripping technology. It could be argued that increased water availability, real or perceived, has enhanced irrigation efficiency, 'saved water' having been employed by most farmers to subsequently expand their irrigated area. This has happened without any adequate external control, neither by CAP regulations, which do not cover these productions, nor by environmental legislation.

The role of the cooperatives in influencing irrigation farmers' decision-making is particularly clear on the basis of our results (an issue we will return to in Chapter 9). Nevertheless, irrigation dynamics result from a complex mixture of water and agricultural policies favouring irrigation farming, which operate in a favourable market for irrigation products. Most farmers argued that the reasons for the changes they made were related to market opportunities (in the case of crop choices) and water saving (in the case of irrigation techniques), which can be summarised as increased profitability - rather than the policy environment - as the main driving force for change. As experts have argued, further irrigation expansion can be expected on the basis of farmers' economic rationality, even under the scenario of full implementation of the EU Water Framework Directive (Escartín and Santafé, 2001). The high marginal productivity of water allotted to intensive irrigation (0.3-0.6 €/m^3; MAPA, 2002) would compensate the future cost of transferable water under the NHP (roughly estimated at 0.34 €/m^3; MIMAM, 2001a) and even of that obtained from seawater desalination plants (Sumpsi *et al.*, 1998). Solutions to the structural water deficit in the basin are, therefore, to be found through managing the *demand* instead of the *supply* of water. For instance, if no further financial resources were provided in the near future, already existing salinisation problems would be exacerbated, lowering productivity and possibly deterring the expanding trend. Although only foreseeable in the longer term in the context of further international trade liberalisation, a possible saturation of European fresh vegetable and fruits markets could also be expected following increased imports from third countries (e.g. Morocco). However, expectations of further availability (either water transfers or desalinated seawater) are encouraging farmers to extend irrigated areas even further. This means that water pricing remains the only available mechanism to control further expansion, an option that would limit water availability to those farms with higher added-value productions (Sumpsi *et al.*, 1998; Albiac *et al.*, 2002). In any case, stricter environmental policy should be enforced in order to control the situation, assuming the foreseeable high economic and political costs of further irrigation expansion.

The situation in the drylands is very different. Dryland farming is an example of a part-time activity, with clear structural limitations deriving from small farm sizes. Under the low cereal cropping productivity in this semi-arid environment, most farmers rely on production diversification based on permanent crops, mainly almonds, olives and vineyards, which have more competitive yields than cereals. In this context, analysed policies have played a contradictory role in the past, simultaneously promoting agricultural set-aside and landuse intensification, while erosion mitigation has never been an objective of any agricultural subsidies. Even rural development programmes have rarely financed farm improvements related to soil conservation, based on the lack of foreseeable profitability increases in the short term. Arguably the most important result relates to the fact that available CAP subsidies, including agri-environmental schemes, have *contributed to the maintenance of marginal farming activities in the drylands*, severely threatened by land abandonment because of low farm-income dependency. This has led to the, at least partial, maintenance of desertification-mitigating farming practices such as terracing or ploughing along contours - crucial management techniques in terms of erosion prevention in the drylands.

Clearer options to focus policies on desertification are starting to appear at present. The new agri-environmental package 2002-2006, with its new measure aimed at mitigating erosion in drylands, is expected to have the most success, given the announced end of subsidies for almond plantations. Further, the inclusion of compulsory compliance with good agricultural and environmental conditions (GAEC) under the mid-term review of the CAP opens new expectations for combating desertification problems in the drylands. However, two conflicting aspects emerge. Considering both the high erosion risk of the environment where these instruments will be implemented and that agri-environmental requirements should go beyond GAEC requirements, means that site-specific needs of protection - which vary depending on degree of slope - should be carefully taken into account when specifying respective requirements. Further, since the cross-compliance approach ignores other factors affecting adoption of conservation practices by farmers (Calatrava, 2004), it may be expected that compliance with GAEC will impose larger burdens on smaller and part-time marginal farms than on more profitable ones, pushing the former further to abandon their farms. Taking into account that, as we have seen in our survey and as has been reported elsewhere (e.g. Wilson and Hart, 2000), adoption of agri-environmental schemes is often influenced by farm size, it becomes clear that larger farms will also tend to benefit disproportionately from agri-environmental support. Therefore, the specific erosion problems linked to *small farms*, either due to redeployment or abandonment, deserve special attention in the future.

A final remark must be made with regard to the lack of reliable available regional official statistics on landuse and crop types, as discussed in our analysis above. Time series on changing landuses and crops in affected municipalities would have been useful for contextualised interpretation of farmers' answers, and for overcoming the deficiencies of our farm survey. Unfortunately, this exercise was not possible given that, due to either mistakes during field data collection or to intentional manipulation, available statistics mask irrigation expansion in the area and appear to offer a contradictory picture of landuse and crop evolution over the past 20 years. It is particularly in view of the conflicting actor spaces where economic development and conservation concerns meet, that accurate and transparent statistical information is most needed. Indeed, design and successful implementation of future policy efforts to combat desertification will rely heavily on such data time series collected by the public administration.

How about
the non-human
actors?

5. Desertification policies in Italy: new pressures on land and 'desertification' as rural-urban migration

A. Povellato and D. Ferraretto

> From here, the landscape was the least picturesque I had ever seen: and because of it I
> liked it very much. There was no tree, no bush, no rock posing as a definite gesture. There
> are no gestures, here, nor the amiable rhetoric of generating nature or human labour. Just
> an even expanse of forsaken land, and the white village above. On the grey sky, a tiny
> low cloud, above the houses, had the vague shape of an angel (Levi, 1945, 23)[23].

This chapter will analyse both the impacts of past policies on desertification processes and
the role of institutional networks and power relations in exacerbating or mitigating
desertification in the Agri basin (located in the Basilicata region of southern Italy). As
Chapter 3 highlighted, this area is one of the worst affected by desertification in the whole
of Italy and has, as a result, formed the focus of international desertification research since
the early 1990s. However, while much information is now available on the physical processes
driving desertification in the Agri basin (e.g. Clarke and Rendell, 2000; Geeson *et al.*, 2002),
there is little information about anthropogenic and socio-economic drivers of desertification.
As with other case study chapters in this book, our analysis will focus specifically on the
impact of policies on desertification, both with regard to policy as a mitigator and driver of
desertification processes.

Section 5.1 will provide a brief description of the case study area to set the scene for
understanding desertification issues discussed in Section 5.2, where the focus will be on
issues of desertification and depopulation and on stakeholder perceptions of desertification.
Section 5.3 will analyse the network of institutions and the most relevant power relations
operating in the area as a basis for understanding the importance of actor networks in the
process of exacerbating and combating desertification - issues discussed in more detail in
Chapter 9. The bulk of Chapter 5 will focus on discussing the relationship between
desertification and policies (Section 5.4), especially by analysing the anachronistic role of
CAP subsidy policies as key drivers of desertification, both in Italy as a whole and more
specifically in our case study of the Agri basin. Concluding remarks will be provided in
Section 5.5.

---

[23] The author, confined in a small village of the Val d'Agri as a political prisoner because of his
uncompromising opposition to Fascism, describes the life of poor peasants in a barren, desolate and
malarial land.

5.1 The Italian case study area

The Agri basin, located in the heart of Basilicata region, is situated north-west to south-east along the 136 km long Agri river with a surface of about 173,000 hectares (Figure 5.1). The Agri basin is divided into three distinct sub-areas (upper, middle and lower) based on physical and environmental characteristics, socio-economic and demographic factors, and based on patterns of human settlements and landuse.

The upper part of the Agri basin is characterised by steep areas and cultivated terraces. It is heavily wooded and mostly covered in thick vegetation that prevents and reduces the risk of erosion. As a result, landslides are infrequent and localised (e.g. recently affecting the town of Montemurro). Agriculture is characterised by specialised crops such as fruit orchards and maize. In the middle Agri basin, water erosion leading to gully formation and the development of 'badlands' is much more pronounced. This area is geologically dominated by flysch and sandstones that are relatively vulnerable to erosion. However, erosion processes in the middle Agri are not solely due to physical factors, such as easily erodible soils (mainly clays and marls), but are also largely due to deforestation over the past century. Land instability is particularly evident during short periods of heavy and strong rain. In terms of landuse, the hilly farmland in the middle Agri comprises durum wheat, pastures, chestnut plantations and vineyards, while olive trees are grown in the lower parts. The lower Agri basin, comprising the coastal area of Metapontino is characterised by stable flat soils. Here, soil degradation processes are largely restricted to riverbanks and to steep coastal cliffs, but salinisation of groundwater is increasingly becoming a problem. Agriculture in the lower Agri

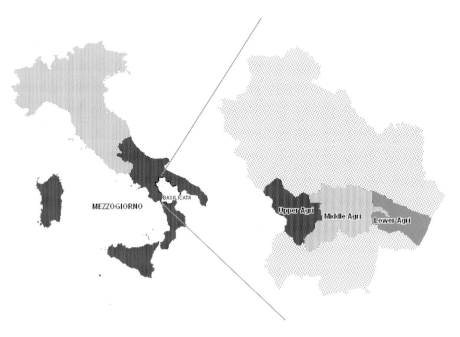

Figure 5.1: Location of the Agri basin case study area (Source: authors).

is characterised by specialised fruit and vegetable cultivation, particularly citrus fruits and horticulture.

As described in Basso *et al.* (1998), population in the middle Basin has decreased, resulting in a low population density of only 32 inhabitants/km^2, while it has increased in the lower Basin to 72 people/km^2. These figures are lower than the Italian average (190 inhabitants/km^2), but are in line with the average for the Basilicata region (65 people/km^2). As will be discussed in detail below, the main driving force behind depopulation is outmigration with continuous population loss in the Agri valley during the 20th century. As a consequence, population loss is a serious issue for most of the basin - not only in socio-economic and cultural terms, but also with regard to maintenance of traditional land management practices that have helped mitigate desertification processes.

The nature of the population problem varies nonetheless between the three regions of the Agri basin. The 'elderliness index', calculated as the ratio between the resident population over 65 and the population under 6 years of age, has different values in the three sub-areas, ranging from an extreme of more than 1,100 in the town of Cirigliano (middle Agri) to 93 at Policoro in the lower Agri. This shows that many villages in the middle Agri are almost entirely inhabited by elderly people, with severe repercussions for how the surrounding landscape is managed due to a shortage of locally available workforce. In addition, the unemployment rate for the entire valley was 30% in 1991, almost three times higher than the national rate. Almost one third of the working population living in the Agri valley is employed in the agricultural sector, with another third working in the industrial sector, which has been recently strengthened by the discovery of oil in the upper Agri. However, there is increasing concern among the local population over the conflicts between preserving natural resources, agricultural land and oil extraction. Indeed, 30 villages of the Agri basin are situated on one of the biggest oil fields discovered anywhere in the world during the past five years, which may herald the end of these village communities when oil extraction begins.

Before discussing desertification issues in the area in detail in the next section, a few comments need to be made about methodologies employed in the data collection for our study (see also Chapter 3). Results will be based on 28 interviews held with policy stakeholders at national, regional and local level, and on a farm questionnaire delivered to 106 farmers in the area. Policy stakeholders were classified into three groups: first, stakeholders who have 'real' decision-making power, including policy makers, actors at the 'policy-decision level' and actors who are responsible for private organisations at a regional level (i.e. farmers' unions) and who have strong connections with public institutions; second, public institution officials responsible for the planning process and policy implementation at the so-called 'design, planning and management level', as well as specialists and technical assistants who operate in public institutions to support policy implementation at the 'technical level'; third, representatives of NGOs at the 'participation level', independent specialists operating in research centres, and stakeholders without any specific institutional affiliation. The farm survey sample was based on a representative cross-section of farms in the upper, middle and lower parts of the Agri Basin, although the sampling strategy focused more on medium-sized and large farms than on smallholdings to ensure that full-time farms, which farm most of the UAA in the Agri basin, were well represented, especially as landuse decisions on these farms are particularly crucial for desertification-related issues.

5.2 Desertification issues

5.2.1 Natural and anthropogenic drivers of desertification

Italy is particularly susceptible to soil degradation due to edaphic, morphological and climatic factors. For a long time, the complex interactions of these factors have prevented the spread of agriculture into some of the most difficult terrains. Anthropogenic activities have, however, left indelible traces on the land over the last millennia due to the gradual spread of population and settlements over most of the Italian peninsula, despite of the fact that 77% of Italian territory is classified as 'hilly' or 'mountainous'. As a result, the landscape has been constantly modified, and natural vegetation has been gradually replaced in most places by new landuses and vegetation better suited to local people's needs (buildings, facilities, food and fibre production). Ever since human encroachment on the land, Italy has witnessed a relatively unstable balance between society and nature, especially as hydro-geological risks are endemic due to the specific climatic, edaphic and structural characteristics of the Italian territory. Nonetheless, soil erosion and land degradation are long-term processes and are only partly due to anthropogenic activities (Gisotti and Benedini, 2000). Thus, water-related drivers of land degradation are partly due to natural conditions of the land, although, as discussed below, water-induced erosion problems are also increasingly linked to human activities. However, it should also be noted that even unfavourable geological and climatic factors can, to some extent, be mitigated by the constant presence of sustainable farming practices which can lead to improvement of soil structure.

Deforestation has been a key driver of Italy's desertification problems. Since Roman times, the replacement of woodland with agricultural land, necessary to provide for the needs of a growing population, has made steep slopes and mountain areas increasingly vulnerable to erosion. Only the constant maintenance of water courses and the use of non-intensive farming practices have made it possible to control land degradation in many deforested areas. However, with the increasing complexity of economic and social systems linked to large-scale land reclamation projects that have occurred since the 19th century, human activities have become increasingly important drivers of desertification. It is also becoming evident that the very socio-economic structures that may have once prevented erosion in vulnerable agricultural areas (i.e. traditional agricultural practices) are currently disintegrating due to recent depopulation trends, thereby often exacerbating desertification.

At national level, polarisation between the lowlands, characterised by rapid industrial/housing development and intensive farming, and the uplands, where land abandonment processes are rampant, has increased since the beginning of the 20th century with the take-off of economic development. As Table 5.1 shows, during the past century the UAA in Italy has progressively decreased from 20.7 million to only 13 million hectares. The dramatic changes in agricultural landuse were characterised by two different trends: on the

Table 5.1: Landuse change in Italy (Source: authors).

Year	Population (million)	GNP per capita 1990 (in euro 1990)	Employment in agriculture (%)	Agriculture	Arable and permanent crops	Meadows/ pastures	Non-utilised areas	Forests/ shrubs	Urban areas	'Sterile' land	Total landuse cover
				Total land cover (1000 ha)							
					of which:						
1910	36.7	1,423	58.4	20,773	15,193	5,580	1,035	4,564	611	1,003	27,986
1929	40.6	1,549	51.7	20,586	14,873	5,713	1,832	5,295	639	1,054	29,406
1955	48.9	2,359	38.8	20,908	15,760	5,148	1,110	5,761	727	895	29,401
1965	52.6	3,397	25.8	20,456	15,320	5,136	1,011	6,089	1,131	766	29,453
1975	55.4	6,366	15.3	17,527	12,313	5,214	3,220	6,309	1,655	754	29,465
1985	57.1	8,158	11.1	16,776	11,800	4,976	2,015	7,785	2,130	757	29,463
1995	57.3	12,447	7.4	16,601	12,042	4,559	1,837	7,660	2,605	760	29,463
2000	57.8	13,540	5.8	13,213	9,798	3,414	1,684	10,967	2,840	760	29,463

one hand, the loss of agricultural land to urban development, particularly in more fertile areas, and, on the other hand, the abandonment of marginal land. Economic development and increasing trade, as well as access to welfare for wider parts of society, are the key driving forces behind these changes (Clark, 1974; Ferro, 1988; Merlo, 1991). This has been associated with a pronounced increase in infrastructural development, which has also greatly affected the above-mentioned fragile hydro-geological balance in mountainous and hilly areas.

5.2.2 Demographic dynamics in southern Italy

Contrasting views about the extent and impact of desertification on the Italian landscape by different stakeholders and stakeholder groups are evident, although most argue that there is a clear link between *demographic dynamics* and desertification processes. This 'human dimension' of desertification is particularly clear in the case of southern Italy, where most human activities have occurred without much coordination (Formica, 1975; Rossi and Iannetta, 2002). Depopulation - linked to anthropocentric definitions of desertification (see Chapter 9) - is caused by lack of proper social and development policy and is a key additional explanation for desertification processes in Italy, together with the lack of a coherent landuse policy framework as discussed below. As a result, many local stakeholders interviewed for our study emphasised the human dimension of desertification, linked to processes of depopulation which characterise most marginal areas in southern Italy and which have become more acute in the last decades of the 20th century. However, the relationship between desertification and depopulation, as well as poverty and desertification, is not straightforward, as the relationship 'poverty-depopulation' in Southern

Italy is usually related to complex social dynamics (Arcieri *et al.*, 2002). Beyond this, it should also be emphasised that some researchers and stakeholders believe that desertification is not necessarily a definitive and unsolvable problem (e.g. Cozza, 2000).

Apart from demographic processes, desertification in southern Italy also has to be understood in the context of other key drivers, especially specific land management practices such as agro-forestry. In areas of southern Italy affected by desertification, the agro-forestry crop system holds a dual role in land management. On the one hand, it may exacerbate soil degradation processes through intensification of agricultural processes and the clearing of trees while, on the other hand, this system can also lead to sustainable environmental management and contribute considerably towards mitigating effects of hydro-geological and climatic factors which often trigger desertification problems. Yet, our research shows that it would be a mistake attributing the responsibility for desertification processes almost exclusively to land managers, without considering the manifold interrelations between these systems and the desertification-affected territory. As our discussion later in this chapter will emphasise, agro-forestry is not a closed system, and the various interactions of this system with other drivers of desertification have to be carefully analysed, in particular with regard to other anthropogenic drivers of desertification such as depopulation.

The impact on natural resource management systems linked to changes in socio-economic structures is directly related to past and recent demographic trends in southern Italy characterised by two distinct migratory processes. First, southern Italy has witnessed 'traditional' migration caused by unemployed southern workers searching employment in northern Italy and, second, by 'modern' migration characterised by more complex patterns that are at times contrasting and overlapping. Traditional large-scale migration, which occurred mainly between the 1950s and the 1960s, was linked to economic and social policy choices made in the early post-war period which tended to play down the unemployment problem in the south and saw it as an opportunity for offering new labour for the rapidly developing areas in northern Italy (Graziani, 1998). This migration was led by workers from the farming sector from southern interior areas (e.g. our Agri basin case study area). It was partly temporary in nature with workers aiming to supplement family incomes by working abroad for a few months a year, but was also characterised by permanent migrants, with up to four million workers leaving the south to find permanent employment and residence elsewhere (Gorgoni, 1980). As a result, many small farms that could no longer survive economically became part-time farms, but were surviving without concrete development opportunities because of lack of capital and because of the rigidity of the land transfer market. It was especially in the 1950s and 1960s that the precarious balance between population and land, built up in the course of centuries, was disturbed dramatically due to large migratory flows pushing millions of people from rural areas in southern Italy to the more developed northern regions. This rural exodus, combined with the abandonment of the most marginal rural lands (see Table 5.1 above), accelerated in the 1960s. Linked to this, southern Italy saw the gradual abandonment of traditional (and often environmentally sustainable) agro-forestry management systems - a process linked to lack of income alternatives for agricultural development and increasing lack of workforce, which both made managing agro-forestry systems less and less profitable.

Recently, 'traditional' migration has diminished and has been replaced by more 'modern' and more differentiated migratory movements. Rural-urban migratory trends continue, motivated in most cases by the search for better living conditions than by the necessity of getting closer to a more dynamic labour market. The scale of these migrations is rather small, and depopulation of small hill or mountain villages is largely driven by lack of access to modern housing and facilities in these areas, with migrants often moving to nearby settlements. Although these migrants form only a small proportion of the overall rural exodus, these internal migration flows risk deeply affecting the socio-economic vitality in hill or mountain areas already subject to large-scale land abandonment. Indeed, below a certain threshold number of inhabitants, the supply of facilities and amenities is no longer profitable, and publicly-funded facilities (e.g. public transport, schools) have increasingly higher per-capita costs. Most importantly for our study, in areas characterised by such upland-lowland migration, the reduced human presence in the uplands also often means decreasing concern towards traditional land management practices, with, at times, loss of desertification-mitigating activities such as terracing and sustainable agro-forestry systems.

Another migratory pattern is also increasingly becoming apparent in the Agri with the growth of coastal tourism. This further encourages temporary migration of workers from areas in the interior of the Agri to the coast during certain periods of the year. The new tourist centres on the coast are also themselves creating increasingly apparent environmental impacts, not only in tourist areas but also in neighbouring areas, particularly by increasing competition for water resources (see below). In the last decade, this competition has increased in southern Italy, and further growth in the tourism sector is expected in the next decades. Nevertheless, tourism can also be linked to improved environmental quality by putting pressure on local decision-makers to enhance protection of local natural and landscape resources (e.g. through creation of protected areas). Thus, tourism, although partly driving desertification processes (especially through water management issues), should also be seen as an opportunity for more sustainable land management.

5.2.3 Desertification and depopulation issues in the Agri basin

From the beginning of the 1980s, policy changes at both EU and national levels have enabled the regional government of Basilicata to implement new policies for the agricultural sector. In particular, the regional government has encouraged young farmers to increase cooperation among landholders through the establishment of fruit orchards, promotion of traditional arable crops, expansion of irrigation and mechanisation, and husbandry of sheep and goats aimed at dairy products typical for the region. As a result of these policies, substantial landuse changes have occurred in the Agri basin over the last decade (Basso *et al.*, 1998). Although beneficial in terms of short-term socio-economic development, some of the resulting changes have directly and indirectly exacerbated land degradation. Thus, trends towards intensification in agriculture and expansion of irrigation have been particularly obvious in the middle and lower parts of the Agri basin, while in the upper part of the basin intensification has been less pronounced as agricultural activities tend to complement other sources of income (see above).

Following typical patterns of agricultural intensification in Mediterranean countries, the Agri basin has also witnessed the transfer of profitable agricultural activities from the uplands to the lowlands, resulting in the marginalisation of agricultural land. In particular, irrigation expansion does not seem feasible in the long run in most areas of the Agri because future water demands for agriculture will not be satisfied by existing reservoirs (i.e. Pertusillo and Gannano reservoirs), which also supply increasing demand from urban and industrial areas in neighbouring regions, and which are already silting up. Establishment of new reservoirs is unlikely, not least because both downstream aquifer recharge is decreased by existing reservoirs and because groundwater contained in coastal sediments is not being replenished (Boenzi and Giura Longo, 1994).

Initial evaluations on the possible choices of public policies to preserve natural resources in the Agri Basin can be found in Basso *et al.* (1998). According to the authors, current population trends in the upper Agri are characterised by consolidation of the existing population in urban centres, while mountain areas are re-afforested. In this area, agriculture is likely to continue to dominate socio-economic development in conjunction with an emergent tourist industry. The latter will be encouraged by the formation of the *Val d'Agri and Lagonegrese National Park*. One strategy for the growth of agriculture is the expansion of horticulture using farm machinery and practices that will not damage environmental and water resources. Regulation of animal husbandry should be used to enhance quality and to diversify production, and traditional agricultural activities in the area could be complemented by the cultivation of fruits and berries. Moreover, marginal and abandoned mountain lands could be returned to agriculture by promoting organic farming of local typical crops, such as high mountain potatoes and pulses. Sustainable environmental management in mountain areas, however, relies heavily on finding the right balance between forestry and grazing. According to Basso *et al.* (1998), development strategies for the middle Agri are more difficult to find because of increasing desertification in the badlands. Here, the most urgent solution is the revegetation of slopes, based on adjustments of existing soil-water regimes, more sustainable farming practices, and the use of more suitable species for pasture improvement. In the lower Agri, meanwhile, we find the almost opposite situation. In the past fifty years, coastal areas have suffered from 'littoralisation' after implementation of large-scale land reclamation projects which led to large-scale population increases (+52% between 1951 and 1961 alone) which has continued unabated until well into the 1990s. At the same time, the middle Agri basin suffered from large-scale depopulation which is continuing to the present day (Table 5.2).

Table 5.2: Migration rates in the Agri basin (%) (Source: authors).

	1950s	1960s	1970s	1980s	1990s
Upper Agri	-0.8	-7.5	+2.2	+0.7	-2.3
Middle Agri	-0.4	-14.5	-10.0	-5.2	-13.2
Lower Agri	+52.0	+10.4	+15	+6.4	-1.5
Total	+9.3	-5.7	+1.5	+0.7	-5.4

As will be discussed in more detail in Section 5.4, subsidy-based agricultural policies linked to the CAP have discouraged farmers from sustainably managing natural resources and, simultaneously, have prevented the adaptation of local farming strategies to current market conditions. This suggests that an irreversible process has started that prevents farmers from adopting 'good farming practice' that would be beneficial for sustainable soil management. Paradoxically, income subsidies and agri-environmental incentives have now become essential for the continuation of agricultural production, both in the EU as a whole (cf. Buller et al., 2000) and in our Agri basin case study area. However, if agricultural activities are reduced or disappear altogether, desertification processes are most likely to further increase. In the following, we will briefly investigate the key farm characteristics in the area and how these influence environmental management decisions on farms.

5.2.4 The influence of farm structural, social and economic factors in the Agri basin

Different from other case studies carried out in various parts of Italy and especially in the south (Eboli, 1992), the Agri basin is characterised by what could be termed 'inverse' off-farm activity, with substantial impact on farm decision-making strategies. In the middle Agri, for example, farm income in 19% of all households is used as economic support for stakeholders involved in other activities than agriculture, especially in the services/production sector (Cioffi, 1997; Ferraretto et al., 2003). Indeed, pluriactivity is considered a basic social characteristic typical of many small Italian family holdings (Marini, 1995) and is no longer considered a transitional phenomenon (Saraceno, 1994). Together with the increase of average UAA on holdings - largely linked to increases in rented farmland - it can be seen to form the driving force for a new food production and farming model (the so-called 'European model') (De Stefano, 2003). Yet, some authors underline the negative impacts of high availability of public employment on those who are considering to become farmers (Stame, 2001) and on local development issues more generally (Trigilia, 1994). Indeed, farmers' decision-making processes with regard to labour-intensive sustainable management practices have been shown to be negatively influenced by too many off-farm work opportunities (Marini, 1995).

It should be emphasised, however, that in the middle Agri more than two thirds of the UAA is still occupied by full-time farmers for whom farming activities make up most, if not all, of the total family income. The size of these farms lies mainly between 20-30 ha (20%), 50-100 ha (20%), and in the 100+ ha category (31%). Freehold tenure remains the most common form of land ownership (34%), even though there are also many tenant farmers (around 13% in the middle Agri). There is some evidence for farm succession, although often dual partnerships between father/son are common, not only because of the relative inexperience of the latter, but also to get EU subsidies paid to young farmers if they directly manage the farm. Thus, often the son is formally in charge of the farm holding, but the real owner and decision-maker is still the father.

These economic and structural factors have major repercussions for sustainable environmental management and land degradation mitigation practices. First, *farm size* has an influence on both income and how the land is managed, with larger holdings depending more heavily on agricultural activity and agricultural subsidies. This means that, in the case of the Agri case study area, effects of agricultural and agri-environmental policy on the environment are linked to farm size, and the larger the farm the more important the policy framework will be for exacerbating or mitigating desertification processes. Second, *younger farmers* are mainly in charge of large holdings, and it is usually these younger farmers who are also better educated. This means that these farmers often show a better understanding of the outcomes of their activities on the environment. Both these key factors are interconnected, which means that (in the middle Agri at least) subsidies are crucial for farm survival on large farms and also strongly influence decision-making processes of the often younger owners of these farms. Thus, often these farmers are torn between having to intensify production in order to gain more subsidies, while having the knowledge that this intensification may further exacerbate desertification processes on their already environmentally vulnerable farms.

Small holdings usually have a lower impact on desertification processes in the area, although this is not necessarily linked to active implementation of desertification-mitigating practices but rather to 'non-action' linked to relatively low subsidy incentives to intensify production, and to the above-mentioned off-farm activities that often mean that these smaller farmers do not necessarily have to maximise production from their farms.

5.2.5 Perceptions of desertification in the Agri basin

As in the Spanish case study area (see Chapter 4), a key aim of our research was also to identify stakeholder perceptions of desertification in order to understand and contextualise policy solutions suggested by different stakeholder groups (see also Chapters 8 and 9). Our first finding from in-depth interviews conducted in the Agri basin highlights that local stakeholders acknowledge that the phenomenon of desertification exists in some form or another in southern Italy. However, national stakeholders were more sceptical about the existence of desertification in Italy than regional and local stakeholders. Second, and similar to findings from the Spanish case study area (see Chapter 4), there is no unanimous consensus as to what 'desertification' means (Arcieri *et al.* 2002). In general, all respondents agreed that desertification was associated with 'land degradation caused by anthropogenic factors', but some national-level stakeholders identified *climate change* as the main driver of desertification - largely based on the level at which they deal with desertification issues (i.e. the national and international scale). On the other hand, local and regional stakeholders viewed desertification almost exclusively in relation to the above-mentioned depopulation problems (i.e. 'human desertification' akin to the French term 'désertification'). Yet, these respondents also largely rejected the relationship between poverty and desertification, arguing that to refer to the Basilicata region as a 'poor' region was not helping to explain the complex dynamics behind land degradation phenomena, and that this was not the 'correct paradigm' with which to ground the political debate. Nonetheless, many other stakeholders, particularly regional actors, saw depopulation as one of the most relevant aspects and causes of desertification (see also Chapter 9).

Interviews with non-farm stakeholders also highlighted that most thought that rural stakeholders' perceptions of desertification were characterised either by not recognising the issue of desertification or by relating desertification processes to relatively simple issues (mainly water availability). On the other hand, the public administration (local and regional) was generally considered to have satisfactory knowledge of the problem. However, our farm survey showed that farmers have a much broader view of the problem of desertification than acknowledged by non-farm stakeholders (Table 5.3). The most common responses by farmers 'correctly' identified almost all of the 'usual' driving forces of desertification (i.e. climatic factors, anthropogenic factors [mainly deforestation] and population dynamics). At the same time, however, and most important in the context of perceived drivers of desertification and associated policy solutions discussed below, there was also a dominant perception that agriculture did *not* play a major role as a possible factor of land degradation (i.e. only 10% of respondents in the farm survey sample acknowledged the importance of 'unsuitable agricultural practices').

Table 5.3: Perceptions of desertification in the Agri basin (%) (Source: Italian farm survey).

Do you think desertification in your area depends on:	Agri basin sub-area			Total
	Lower Agri	Middle Agri	Upper Agri	
climate change	52	59	45	56
Deforestation	17	36	0	27
Drought	48	47	27	45
Depopulation	0	42	64	34
Land degradation caused by human factors	4	17	0	12
unsustainable use of water resources	35	0	9	10
crops which are not suited to land	0	2	0	1
unsuitable agricultural practices	30	2	9	10
Other	13	3	0	5

5.3 Institutional networks, power relations and desertification

5.3.1 Actor networks

As part of the actor-oriented approach used in the MEDACTION project (see Chapters 1 and 3 and also Chapters 8 and 9), an important aspect of our in-depth interviews with stakeholders in the Agri basin was to obtain information on the *role* of specific actors and actor groups and their relative *power* with regard to policy implementation networks (Arcieri *et al.*, 2002). This is particularly important in light of our discussion in Section 5.4 on the specific impacts that policies and policy decisions have had on desertification processes in our case study of the Agri basin.

At the national level, three ministries (Ministry of Environment, Ministry of Infrastructure and Transport, Ministry of Agriculture) have partially overlapping roles in planning and management of Italian water and soil resources. However, based on the Italian political structure and the important role of the regions, the central policy decision-making role with regard to our Agri basin case study area is held by the regional government of Basilicata (see Figure 5.1 above). The process of empowerment of the region began with decentralisation in the 1970s (so-called 'regionalisation'), and regional government can today be considered as the main policy actor, especially with regard to agricultural, water and soil management policies. The region is at the centre of a network of national and local bodies, acting as a direct and powerful intermediary by implementing national policies according to the region's needs or by autonomously implementing new policies addressing above-mentioned issues.

Nonetheless, the regional government is not always an independent decision-maker. With regard to agricultural policies, for example, a large number of market policies are implemented at national (and indeed EU) level, with little opportunity for influencing policy directions at regional level. The case of durum wheat subsidies, examined in more detail below, is a particularly obvious example, although the key role played at the local level by farmers' unions is also interesting to note in this respect. Farmers' unions represent the main interlocutor with the regional government on agricultural policy matters, influencing policy implementation by facilitating the submission of applications for subsidies. Thus, as in the Spanish case study discussed in Chapter 4, farmers' unions have played a key role with regard to enabling farmers to obtain maximum subsidies for environmentally damaging intensive durum wheat crops that are increasingly replacing traditional crops, with resulting negative effects for desertification mitigation (Arcieri *et al.*, 2002).

In other policy arenas, however, the regional government has established powerful control over the direction of policies. For example, the region of Basilicata can be criticised for having been responsible for what could be termed a 'distorted' use of social policy incentives designed by national policy makers to improve social and economic conditions in southern Italy. Evidence suggests that, rather than channelling social policy funds to the neediest in society, local administrators used these funds to increase their own power in decision-making networks through the control of large national funds theoretically destined towards improving overall welfare of the area (Trigilia, 1994). As a result, many funds have been, for example, allocated to the local forestry sector to offer employment, especially to people living in marginal areas. Although initially this policy led to some indirect benefits with regard to desertification, especially through positive long-term protection of soils from erosion, direct and indirect negative effects have also increasingly emerged. These have been largely linked to the emergence of 'clientelism' which damaged the local labour market. Farmers in the Agri, in particular, have complained about the difficulty to find workers, as since the 1980s many workers preferred working for the forestry sector as part of an increasingly important 'underground economy' that could command higher wages than the agricultural sector. Thus, interviewees saw the focus of local policy on forestry as a major problem, and saw this as the product of social policy 'patronage' by regional politicians who have focused on the forestry sector to gain political support among populations from the less developed areas in Basilicata. The resulting transfer of 'social' funds from agricultural development towards wages for forestry workers has led to the creation of 'ghost workers' (Anania *et al.*, 1992) - i.e. where money could be claimed from these funds by powerful elites for forestry work that was never conducted. This situation was exacerbated by inadequate budgets for infrastructure, tools and management with regard to newly established forests, a problem which has not been resolved to the present day, despite of new financial resources coming from oil exploitation. This situation has meant that it has been increasingly difficult for farmers in desertification-affected areas to find people available for 'regularly' employed farm labour, resulting in the fact that farm activity has increasingly been dependent on family work. In the long term, this labour shortage on farms is leading to further neglect of traditional labour-intensive sustainable farm management strategies with concurrent negative effects for desertification mitigation.

5.3.2 Policy agendas in the Agri basin

The discussion in the previous section has highlighted how specific actor groups in the Agri basin play a dominant role in regional and local policy decision-making. In this section, we wish to analyse these power relations in more detail by looking specifically at the notion of policy 'agendas' - i.e. policy priorities that individual stakeholder groups have when designing and implementing policies that may affect desertification. These issues will also be explored in further detail in Chapters 8 and 9. Over the past decades, national and regional policies in Italy have pursued the overall goal of developing the Mezzogiorno (of which the region of Basilicata is part). Many policies that affect desertification in the Agri basin have to be understood within the framework of this wider policy goal, and most of the policy agendas described in the following have to be contextualised within the context of the wider policy aim to help 'development' of the region of Basilicata. Four dominant policy agendas can be identified among stakeholder groups in the Agri valley.

First, the *modernist agenda* aims at the development of the local and regional economy, with a particular focus on the most viable and technology-heavy landuses (see also Chapter 9 for further detail on different policy 'agendas'). As mentioned above, the lower Agri has experienced large-scale influx of people from interior areas of the Agri attracted by better standards of living and work opportunities based on large-scale land reclamation programmes in the lower Agri since the 1950s. All subsequent policies focused on technological improvement, with little or no concern for environmental sustainability. The result was the creation of highly productive irrigation areas focused on horticulture. As part of the dominant modernist policy agenda, the regional agricultural plan (2000-2006) for the lower Agri provides further incentives for the establishment of new irrigation plans. Although these are hoped to encourage good management and water savings, evidence suggests that, in reality, this modernist policy agenda leads to an unsustainable economic development model that exacerbates problems of poor water quality and availability because of overexploitation of water resources. An example of the lack of recognition of such mismanagement by those adhering to the modernist agenda is given by a local stakeholder who argued that, in his opinion, salinisation in the lower Agri has not resulted from overuse of groundwater by farmers but from aridity (i.e. lack of surface water). This emphasises the instrumentalist interpretation of the desertification problem by those advocating a modernist policy agenda who apportion blame to 'external' factors rather than to internal agriculturally-related drivers.

Second, what could be termed a *productivist policy agenda* can be identified in the Agri basin, an agenda that aims to maximise economic income from agriculture in the local area. This agenda is particularly evident in the middle Agri, where agriculture represents the sole activity for a large percentage of the population. There, cattle breeding and cereal crops (often in combination) are traditionally the main agricultural activities. In this context, uncritical acceptance of subsidies linked to the CMO for durum wheat has been particularly prominent, resulting in intensification and increased land degradation (see also discussion in Section 5.4). A local stakeholder argued that agricultural intensification in the middle Agri is not seen as damaging by stakeholders adhering to the productivist policy agenda, and that many see durum wheat subsidies as a necessary incentive to keep farmers in the area. Thus, land abandonment, in the opinion of this stakeholder, would have worse effects than the uncritical use of agricultural subsidies leading to intensification. It should also be emphasised that it is difficult for farmers (especially those living in marginal areas) to obtain loans from banks or to be able to afford any substantial farm investments. This has meant that the durum wheat subsidy has often represented the *only* way to obtain some revenue for these farmers, and reduction of these subsidies, in light of Agenda 2000 reforms, is perceived by farmers as the possible end of farming in the middle Agri. It can be argued that increasing dependence of farmers on this income has led to the detachment of this subsidy from its original significance. Indeed, attempts to 'crop the subsidy', rather than the wheat, have characterised the predominant farming philosophy in the middle Agri - with severe repercussions for desertification processes as discussed in more detail below.

Third, the productivist policy agenda has to be interpreted in the context of other policy drivers affecting decision-making in the Agri. In particular, what could be termed the *welfarist policy agenda* - an agenda typical for Italian policy-making - has played a crucial role. Indeed, the peculiar social and economic context of southern Italy may be more easily understood if we take into account the importance of this agenda. As described in Arcieri *et al.* (2002), there is clear evidence of an instrumental use of certain policies to gain political consensus. The most blatant example is that of regional forest policy, which, as discussed above, has provided a framework for the allocation of funds with the primary aim to employ as many workers as possible, even if there are no tools or machines to work with. A study conducted in the upper Agri by Cioffi (1997) argued in this context that the service sector has been the only one to see increases in the number of employees, to an extent that this sector now forms the main activity in the area. Many national policies (e.g. policies aiding industrialisation of southern Italian mountain areas) have been implemented by regional policy-makers in the same way as these forest policies, and it was argued that investments were needed to demonstrate the willingness to help the Mezzogiorno 'by any means', despite of a clear lack of cost-benefit analysis and feasibility studies analysing how to best use funds linked to these policies. The motto, therefore, was 'the more money, the more political consensus', although often little (or none) of that money reached local populations. Trigilia (1994) argued that even if funds trickled down to the grassroots level, it was largely by means of income subsidy. Thus, the predominance of this welfarist agenda, closely linked to the productivist policy agenda highlighted above, creates a social environment with low levels of social capital, thereby increasing the dependence of people both on institutions and top-down economic support, while simultaneously empowering local politicians. While the effects of policies based on the welfarist agenda on desertification processes in the Agri can be positive, as in the case of afforestation, in most cases effects have been negative by exacerbating desertification as exemplified through the case of irrigation policies in the lower Agri highlighted above (see also Section 5.4).

However, several efforts have been made to mitigate policy impacts on desertification in the Agri basin, leading to a reactive *conservationist policy agenda*, the fourth agenda influencing desertification processes in the Agri and the most 'positive' with regard to future prospects for desertification mitigation (see also Chapter 9). This is particularly relevant with regard to water resource management as, since the 1990s, water shortages have worsened due to both increasing demand and reduced availability. This has made local communities more aware of the risks associated with both their economic and land management activities. The Integrated Water Service (established 1996) and the Programme Agreement (established 1999) with the region of Apulia, together with technical assistance provided by the region of Basilicata to farmers, are clear steps towards a new, broader and more sustainable management of local and regional water resources. Nonetheless, as was the case in the Spanish case study discussed in Chapter 4, short-term planning often jeopardises these positive developments, and large volumes of water are still abstracted, based on unsustainable 'water resource balances' calculated by planners adhering to the productivist agenda. This suggests that the reactive conservationist agenda is still far from being fully taken into consideration by policy-makers. In the worst case scenario, this policy agenda could also be appropriated by those stakeholders wishing to gain political consensus (as is the case with the above-mentioned welfarist agenda).

5.4 Desertification and policies

Having outlined desertification issues in the Agri case study area and different policy agendas prevalent among various stakeholder groups, in this last section we wish to analyse in more detail the *interlinkages* between desertification processes and policies. We will do this through a stepwise analysis that starts with the national framework for landuse planning in Italy and then analyses the role of agricultural, forestry and development policies in exacerbating or mitigating desertification processes in the Agri basin.

5.4.1 The national framework for landuse planning and natural resource management

The first policies implemented to reduce environmental degradation began in the early years after the unification of Italy, starting with Law 2248 in 1865 which defined the role of the state in land management issues. However, it was not until the first decades of the 20th century that the first effective land management policies emerged. According to Gisotti and Benedini (2000), three crucial phases can be identified with regard to policy implementation for environmental management in Italy: first, the period between the 1920s and 1930s with implementation of the first systematic land management policies; second, the period between 1970 and 1977 when first steps occurred with regard to transferring technical and administrative powers for policy implementation from the national to the regional level as part of the process of decentralisation; and, third, the 'approval phase' between 1989 and

1994 for two key policies that currently define rules for the protection of land and for water resource management for urban uses.

In the first half of the 20th century, three policies had the potential to positively affect land management in Italy. The first was Royal Decree 3267 in 1923 which identified about half the national territory as vulnerable to environmental degradation, and that attempted to regulate landuse practices that negatively affected water and soil resources - especially by preventing the cutting of protective woodland - with the aim towards improved soil protection. Second, Decree 215 in 1933 put in place basic principles for state intervention in landscape protection issues, acknowledging that Italy's land resource had multiple functions including environmental protection, production and settlement functions. However, during this period, landuse policies, largely based on the above-mentioned productivist policy agenda, still placed a primacy on agricultural production, while soil protection was usually only a secondary goal after food self-sufficiency was guaranteed. This meant that policy mechanisms alone could not guarantee protection from desertification, and that only permanent (and non-intensive) cultivation could limit natural causes of desertification linked to hydrological and geological factors. Indeed, due to the wide availability of low cost farm labour in the first half of the 20th century, traditional agricultural systems could be maintained with more or less success in many desertification-prone areas with often positive effects for soil conservation. The third key policy (on landuse planning) had even less positive impacts. Implemented in 1942, it focused in theory on the crucial issue of landuse planning, but it only dealt with the issue from a strictly urban planning perspective, and only considered agricultural and natural areas as potential sites for housing development, industrial activities and transport infrastructure (Cianferoni *et al.*, 1976; Cannata and Reho, 1981).

The decentralisation of land management responsibilities (urban planning, environmental management and agriculture) to regional administrations, starting in 1970, made coordination of policy mechanisms and impacts even more difficult. This already became evident in the 1970s when a parliamentary survey, initiated after the dramatic floods of 1966, pointed towards the necessity of producing a systematic information base for Italy's national land resources, the drawing up of river basin plans, the reorganisation of national technical advice services, and the increasing allocation of funds for soil protection that should be more substantial and permanent. Yet, only twenty years later (in 1989) were these recommendations implemented through policy 183/1989 on soil protection. This latter policy, based on the reactive conservationist agenda, has provided the legislative framework for environmental preservation. The notion of soil protection enshrined in this policy includes water reclamation, the use of water resources for economic as well as social development, and environmental protection prescriptions. The geographical reference point of the new planning framework, now based on river catchments and no longer on administrative boundaries of local administrations, has to be seen as a very innovative and bold step forward with regard to desertification mitigation.

Further incentives for the development of a normative and more holistic framework for water conservation in Italy was given by EU environmental directives in the 1990s. For example, policy 152/99 on water conservation, based largely on EU directives 91/271 and 91/676, repealed or substantially modified previous legislation on water resources. This policy took into account changes introduced at European level through the EU Water Framework Directive, and introduced an innovative approach towards water conservation (see Ficco *et al.*, 1999, for further detail). Most important in the context of our study, it also made reference to desertification, and took into account the identification of desertification-affected areas based on the Italian NAP (see Section 5.5 for a brief discussion of the most recent policy developments regarding the Italian NAP).

In 1994, the passage of the so-called 'Galli law' (Law 36) completed the regulatory framework, with a number of guiding principles which now have to be taken into account in any planning and management issues regarding water resources. The Galli law works on the principle that water conservation is in the public interest, thereby adhering to the polluter-pays-principle. Unfortunately, and as is often the case in policy-making in Mediterranean countries (Garrido and Moyano, 1996; Buller *et al.*, 2000), these guiding principles have not been fully implemented yet (10 years after the law was passed!), and the resulting shortcomings from this new legislative system seem, at times, to exceed the good intentions that lie behind it. It also seems evident that, although well intended, policy-makers have not yet succeeded in reorganising and devolving decision-making powers away from powerful national and regional elites who continue to drive policy agendas. This continues to make implementation of landuse policies aiming to combat desertification at the local level very difficult.

This situation is made worse by the fact that institutional arrangements for water and land resource management operate at different levels. Thus, the state is in charge of legislation and implementation of EU Directives, while administrative regions are in charge of water and soil resources planning, pollution control, and all administrative issues. Issues of inter-regional interest, meanwhile, are dealt with by river catchment authorities (i.e. operating at yet another scale), while local authorities are in charge of organisation and management of water services. To add a further dimension to this already complex situation, management of land and water resources can also involve corporate institutions (sometimes institutions closely linked to the state such as the *Land Reclamation Syndicates*), who may also choose to work with different levels of the public administration. Massarutto (1999) has argued that this complex framework has led to highly differentiated water and soil management policies at the regional level, with policy implementation negatively affected by a relatively unclear and contradictory normative policy-making framework that includes various uncoordinated authorities. In particular, there are still unresolved issues linked to the relationship between national government and the regional administrations, and our interviewees suggested that currently available operational tools are inadequate to appropriately enforce policies. Finally, problems also emerge due to often poor skills of administrators (Senato della Repubblica, 1998).

Since the 1950s, the evolution of soil conservation policies in southern Italy has particularly depended on the activities carried out by the *Fund for the Development of Southern Italy* (CASMEZ). CASMEZ sub-contracted the task of land reform on large estates that still controlled most of the agricultural land in Italy in the 1950s, as well as land reclamation processes on the coastal plains, to both the *Land Reclamation Syndicates* (LRSs) and the *Land*

Reform Agency to promote the development of rural areas. However, the LRSs were not able to achieve complete land reform because of opposition at local level by large landowners who continued to hold much political power. The LRSs eventually became public work contractor agencies that also had duties external to the farming sector (e.g. civil infrastructure development and road building). As a result, they lost both their role in the management of land reform processes and crucial contacts with their associates. In the 1970s, competencies regarding soil protection planning and management passed to the regional level, which made the LRSs entirely dependent on the regional government. Although regional governments (e.g. Basilicata) maintained the functions of the LRSs, they took away LRS control over general soil protection planning and limited the LRS's role to managing specific situations and emergencies. In this new organisational model, LRSs lost their political influence based on direct election of administration board representatives, and other local bodies such as mountain communities and communes became more important in the regional planning process (Arcieri *et al.*, 2002). In this context, it can be argued that CASMEZ interventions, and the social policies associated with CASMEZ, clearly exhibited features of the above-mentioned welfarist policy agenda, where common interests emerged between local politicians who were often also members of Parliament on the one hand, and potential voters interested in maintaining their subsidies on the other. The above-mentioned case of forestry workers is a classical example of this 'symbiotic' relationship.

In the Basilicata region, the first stage in which CASMEZ paid particular attention to land development was followed by a second less forward-looking stage in the 1970s. A complex special irrigation programme was undertaken in the lower Agri between the 1970s and 1990s, where acquisition of new irrigation equipment was supported well beyond reasonable needs, resulting in extension of irrigation areas without due consideration of actual water availability (see also Chapter 4). After implementation of irrigation measures, land transformation without real planning regarding actual water availability was begun. The imbalance between water availability and area under irrigation was also underestimated in the planning of agricultural development within the Basilicata region operative programme from 1994 to 1999 (part of the structural funds package within EU Objective 1 areas). Some measures in this programme had envisaged the restructuring of horticulture, which would have allowed the expansion of fruit and vegetable crops. Farmers in the lower Agri have indeed responded positively to these incentives and planted new vegetable and fruit crops. However, people are now aware of the fact that the Basilicata region is faced with severe threats of water shortages, and the problem is also extending to availability of drinking water. As a result, methods of water resource planning and management have been increasingly debated among various stakeholder groups, and possible mistakes made in landuse planning are beginning to be acknowledged. A positive outcome mentioned in our interviews is that these re-evaluations may lead to future re-orientation in the management of water resources. Stakeholders increasingly acknowledge that this is needed in order to be able to cope with the reduced availability of water and to limit, as far as possible, the resulting effects on soils, environment and productive activities.

A final set of national policies that needs to be considered with regard to land degradation and desertification processes are policies that focus specifically on conservation issues, in particular national park and landscape conservation policies. However, these policies have limited effects on desertification processes as a whole, as protected areas cover less than 10% of the national territory, targeting only a small, albeit significant, proportion of land threatened by degradation processes. Nonetheless, mandatory landscape conservation schemes cover almost half the national territory, and in 1985 conservation policy was also drastically widened with the passing of a new policy for the conservation of areas of particular natural and cultural interest (Law 431, also known as the 'Galasso Law'). The wide range of protected areas is based strictly on their environmental value, reflecting the emergence of a new vision of landscape and environment no longer purely related to visual and aesthetic attributes of the environmental resources to be protected, but also acknowledging intrinsic environmental values such as biodiversity (Casadei, 1991). Several million hectares of land were protected, including relatively large areas where any modifications in landuse and building styles were prohibited. The law also made implicit recognition of the positive role played by agriculture in landscape formation.

In 1991, the new guidelines on protected areas became law, after twenty years of preparation. The new policy (Law 394/91) includes at least two positive elements for desertification mitigation. First, the general approach of this policy, based on the attempt to combine conservation and enhancement of the natural heritage with the development and promotion of 'environmentally sound' productive activities, has to be seen as a positive step for desertification mitigation (Montresor, 1994; Grillenzoni and Ragazzoni, 1995). Second, with regard to institutional structures a new and positive division of decision-making powers has been established, with a distinction between nationally and regionally protected areas. However, yet again the implementation process is slow, and Moschini (2000) has highlighted that bureaucratic delays, financial difficulties at national and local level, and the lack of political interest in environmental conservation are all resulting in substantial delays in the establishment of new national park authorities. Yet, new policy mechanisms (e.g. Law 426/98) have given a new impulse to improving park policy, especially by assigning a more active role in environmental conservation to local stakeholders, and by emphasising the value of cultural heritage and local traditions (INEA, 1999).

5.4.2 The role of agricultural and forestry policies in natural resource management

In this section, we wish to look specifically at agricultural and forestry policies as potential key drivers of desertification in Italy. This will form an important basis for understanding the differentiated effects of policies on landuse practices in the Agri basin discussed in detail in Section 5.4.3.

Agricultural policies

Until the 1980s, the general focus of Italian agricultural policy was to improve agricultural development and farm structures, to forge permanent links between farmer and farm, and to give priority to policies helping small and medium family farms. Many authors commented on the contradictory goals of agricultural policy before 1980, with some policies focusing on increasing production ('productivism'), while other agricultural policies aimed at improving social welfare in the countryside (Fabiani, 1986; Fanfani, 1990; Guariglia, 1994). Productivist tendencies were especially evident on more dynamic farms and in intensive farming areas, allowing the agricultural sector to increase production significantly. From a regional point-of-view, investment by the private sector was almost exclusively limited to areas of northern and central Italy, while in the south investment in agriculture was generally geared towards public infrastructures (Fabiani, 1986).

The general approach to agricultural policy and to the predominant productivist policy agenda did not undergo any significant changes in the 1990s, and there have only been small adjustments of direction due to recent economic trends. As a result, the gap between production- and welfare-oriented interventions has become even wider. Some attempts were made to formulate a national agricultural policy more suited to the new context in terms of changing agricultural structures and improving farm services and marketing infrastructures. Yet, the problem of the increasingly wider gap between regions was almost entirely overlooked, and most of the expenditure was devoted to more fertile areas. Schemes put in place did not effectively rationalise expenditure on structural incentives, and as there were no definite selection criteria small and large farms alike were beneficiaries of financial subsidies. Yet, in the mid-1980s significant differences began to emerge between small and large farms with important repercussions for land abandonment (see above), although it was argued that the overall objective of income support should be achieved by improvement of *productivity*, rather than by increasing production which had been the basis of agricultural policies up to then. As a result, employment security, reduction of north-south disparities, environmental conservation, reduction of reliance on external imports of food, and incentives for southern Italy were all objectives/constraints in the *National Agricultural Plan* approved in 1985. In line with parallel developments at the EU level (cf. Buller *et al.*, 2000), this was the first time that adequate attempts were made in agricultural policy to address environmental problems.

Implementation of various CAP reforms between the 1980s and the 1990s found most Italian decision-makers generally unprepared. The problem of surpluses, the need of balancing the ratio between internal and international prices for some products, and environmental protection issues for the countryside, were all not on the domestic agenda during that period, and even in the 1990s attempts to reform national agricultural policy were half-hearted and poorly implemented. As a result, provisional and short-term legislation was approved without substantial modification of previous productivist policies. The policy decision-making process has been further constrained by an increasingly tight public budget, leading to a significant reduction of both state expenditure for agriculture in general and new schemes implemented as part of the CAP reform in particular.

One of the key policy drivers for landuse change in southern Italy has been the implementation of regulations supporting durum wheat crops (Regulations 1765/92 and 1253/99). This has led to an uncontrolled increase in durum wheat cultivation, with concurrent detrimental effects on natural ecosystems. The reason for large-scale soil degradation and ecological damage created by this 'distorted' implementation of EU regulations is particularly linked to erosion-prone soil types prevalent in many areas of southern Italy that react poorly to more intensive cultivation (Phillips, C., 1998; Clarke and Rendell, 2000). The situation was made worse because the submission of applications for durum wheat subsidies was organised almost exclusively by farmers' unions, and as these unions are paid by farmers for their services, this mechanism has tended to further increase the number of applicants receiving durum wheat subsidies without consideration of the overall farm structural and environmental capacities of these rural areas. This has been a particularly unfortunate trend with regard to desertification mitigation, as in recent times many farmers, worried about continuing yield decreases, had begun using more environmentally friendly farming techniques such as more shallow ploughing (less disturbance to topsoils), sod seeding, or minimum tillage practices - all leading to reduced environmental impact. Even in the heavily desertified *calanchi* areas some farmers tried these new techniques which reduce costs and soil erosion, thereby fitting well with the farmers' aim to obtain a satisfying income. Yet, it is in these areas, in particular, that increased durum wheat planting, sparked by the subsidy regime, has superseded traditional and often more environmentally-friendly crops and farming practices.

Forestry policies

As highlighted above, forestry plantings carried out with CASMEZ funds during the 1950s had a positive impact in terms of soil and slope protection. However, when administrative functions were transferred to the regions in the 1970s, fewer investments went into soil protection initiatives, while, at the same time, the regions ceased implementing policies specifically targeting environmental management in mountain areas. In this context, implementation of afforestation measures serves as a good example of the decreasing trend of CASMEZ investments. Before the 1970s, afforestation was mainly carried out by the *Land Reclamation Syndicates*, by the *Irrigation Development Office* and the *Forestry Office*, all funded by CASMEZ. This enabled the employment of several thousand workers for many months per year. However, when these competences passed to the region, there were insufficient financial resources and afforestation incentives had to be stopped. As highlighted above, forestry workers continued to be employed for 'social purposes' without any requirements for minimum work aimed to restore woods. This distorted situation has been maintained by politicians who have continued bestowing favours by employing as many forestry labourers as possible. This highlights that social aspects prevailed over environmental considerations due to the dominant ethos of favouritism typical for the welfarist policy agenda highlighted in Section 5.3.

In recent years, so-called 'productive reforestation' linked to EU Regulation 2080/92 has been widely carried out using, in particular, species such as cherry, walnut and chestnut. However, such species are ecologically very demanding and therefore not always suitable for marginal soils. Interviewees highlighted that such policies were indeed motivated by opportunities to get both higher premia and priority in project selection, which explains why funding applications were most often for the planting of economically valuable species such as walnut or cherry which produce top-grade timber.

Agri-environmental policies

There is no doubt that, up to now, EU Regulation 2078/92 (the 'Agri-environment Regulation') has been the most relevant policy framework for environmental conservation in rural areas in Italy. The area enrolled into agri-environmental measures reached almost three million ha in 1999 and, although the area enrolled had slightly declined at the time of writing (late 2004), agri-environmental measures included in the *Rural Development Regulation* since 2000 continue to constitute the backbone of Italian agri-environmental policy. Not only have agri-environmental measures created new income opportunities for farmers that form viable alternatives to traditional productivist commodity production, but they have also engendered diffusion of environmentally sound farming practices and the production of agri-environmental goods.

A study recently conducted by INEA (1999) showed that Regulation 2078 has been recognised in Italy as an instrument reducing the negative impact of agriculture on the environment. However, when looking more closely at measures that relate to land management and, therefore, to desertification processes, the notion of countryside stewardship has only been recognised as an essential factor in landuse management in some regional areas (particularly in northern regions), where these objectives had already been part-and-parcel of farming practices for a long time. In the central and southern regions, meanwhile, acceptance of agri-environmental measures has not met expectations, partly because planning and preparation for such new policy types has been lacking. Unfortunately (in desertification terms), this was particularly the case in areas that would have benefited most from introduction of environmentally sound techniques in pasture and extensive meadow management. Similarly, despite of gradual increases in uptake, the measure for 20-year set-aside for environmental purposes showed uptake levels well below initial expectations and was mainly concentrated in the south of Italy. On the other hand, measures helping to prevent farmland abandonment and for the management of forestlands had good uptake levels, especially in the south.

Despite of these partial implementation shortfalls and problems, the importance of Regulation 2078 as a tool to modify both farming systems and farmers' attitudes has to be emphasised - in other words, as a tool potentially shifting productivist policy agendas towards conservationist policy agendas. According to first evaluation results, the implementation of integrated farming, for example, allowed for a reduction of environmental impacts of agriculture in the more intensively cultivated areas, mainly because it includes crop rotation - a practice otherwise no longer used in most Italian farming areas. Most important in the context of desertification mitigation, such crop rotation practices are again becoming prominent in dryland farming areas affected by desertification. Yet, the problem with this type of policy concerns the duration of the agreements, especially as five years is too short a time period to ensure permanent soil conservation. Thus, if there is no chance to renew the agreement, farmers take to intensive cropping again and resume durum wheat monoculture based on the high income margins linked to this type of cultivation (see above). Further, the relatively undemanding changes to farming practices imposed by integrated farming may lead farmers to consider this specific measure as a mere income subsidy. In line with calls from other studies on the effectiveness of agri-environmental policies in the EU (e.g. Buller *et al.*, 2000; Wilson and Hart, 2000), this suggests that better scheme design and implementation of more rigorous control systems will be necessary to ensure maximum environmental benefits (i.e. desertification mitigation) from these measures in the future.

The first results of studies on the effects of Regulation 2078 have highlighted complex patterns in connection with farm types targeted by specific policy mechanisms. The large number of small farms in Italy, for example, emphasises the importance of improved policy targeting, and there are also many part-time farms that are often ignored by the CAP policy framework, notwithstanding the substantial share of land that they manage. Studies show that small full-time farms, in particular, have efficiency problems because of the greater difficulty in adapting to economic changes and social problems, largely due to limited family income almost exclusively derived from farming. As Cafiero *et al.* (1998) have emphasised, these issues can greatly vary according to the features of the local economy and market. Some authors have, therefore, stressed that Regulation 2078 could be used to counteract negative effects of direct payments for arable land in mountainous and hilly areas (e.g. Dono, 1996). Indeed, current direct payments to these farmers lead to an increase in the area of arable land and the gradual disappearance of traditional farming types based on small-sized livestock farms and perennial green fodder crops, with concurrent negative effects on the environment due to loss of soil protection cover, shorter crop rotations, and lack of organic fertilisers.

Finally, the connection between advisory services and agri-environmental measures also needs to be considered. Poor interlinkages between farmers and advisory services often lead to organisational problems about modalities and types of services offered. Where the 'food chain approach' prevails among advisory services (i.e. services are delivered on a single-production base and advisory staff are specialised), the farmer is likely to be less easily convinced to enrol in an agri-environmental scheme that would imply changes in farm management with reference to the entire farm area. However, if the extension service is based on advisory staff who focus on the management of the whole farm, promotion of agri-environmental measures may lead to better uptake results. Moreover, the direct involvement of extension services in the implementation of agri-environment schemes (e.g. linking the delivery of public funds to advisory services specifically for conservation efforts) may give greater impulse in contacting new farmers and widening the range of services supplies. Further, applications by farmers for different measures in the agri-environmental package may be strongly influenced by transaction cost assessments conducted by advisory staff, thereby encouraging application for 'easier' measures and reducing advisory efforts for farmers to apply to more complex measures. Yet, it is the latter types of measures that are often most beneficial for desertification mitigation.

5.4.3 Policies and desertification in the Agri basin

Having outlined the general framework for understanding the linkages between policy and desertification for the whole of Italy in the previous two sub-sections, in this last part of our analysis we wish to focus on the impacts of these policies on landuse practices in our case study region of the Agri basin. First, we will look at the key phases of policy-making in the Agri and, second, at the specific impacts that agricultural polices have had for encouraging or discouraging environmentally-detrimental landuse changes.

Key policy phases in the Agri basin

Policy effects on landuse in the region of Basilicata (and in the Mezzogiorno overall) fall into three distinctive periods (Arcieri *et al.*, 2002). First, the 1950s and 1960s were characterised by large-scale land development programmes (land reclamation, construction of dams, creation of new settlements in coastal areas) supported by above-mentioned CASMEZ funds. A result of these programmes was that the difference of income per capita between southern and northern Italy largely remained the same between 1950 and 2000, suggesting that the Mezzogiorno has kept pace with one of the most economically dynamic areas of Europe (Trigilia, 1994). The 1950s and 1960s, in particular, emerge as a period where extreme poverty was eradicated from most rural areas in southern Italy (De Benedictis, 2002).

Second, during the 1970s and 1980s, policies supporting further land transformation were introduced, funded through the *Fund for Investment and Employment* and consisting mainly of large-scale infrastructural works. However, despite substantial infrastructural developments including considerably improved networks of main roads, secondary roads in the countryside continued to be in poor condition as they were planned and built before cars became the dominant mode of transport. As Table 5.4 shows, respondents in our farm survey highlighted that poor infrastructure in central parts of the Agri basin was the second most important factor negatively influencing farm survival in the area. The lack of a proper road network was seen as further exacerbating isolation from the local community, making everyday family life (e.g. reaching school and services) difficult.

The third policy phase occurred during the 1990s when the state increased financial resources to help sustain incomes in the Mezzogiorno. However, funds linked to this policy phase focused largely on social policy and administrative expenses, while rural development programmes received less than 10% of the total amount - in contrast to the average 15% of funds obtained by farmers in northern Italy (Trigilia, 1994). As a result, overall per capita

Table 5.4: Importance of public services perceived as essential by farmers (%) (Source: Italian farm survey).

Which of the following public services will help financial survival of your farm?	Agri basin			Total
	Lower Agri	Middle Agri	Upper Agri	
More and easier access to credit	24	6	6	10
Strengthened cooperatives and producers' associations	12	13	18	13
Make access to EU funding easier	12	9	35	14
More technical aid to farmers	20	5	24	11
Training and professional upgrading	32	5	6	11
Improved public services (roads, transport, etc.)	12	61	0	40
Offer new marketing opportunities for farm products	60	72	88	72
Other	12	0	6	4

income in the Mezzogiorno declined from 60% of the northern Italian average in the early 1980s to 57% in 1999 (Rossi-Doria, 2003). De Benedictis (2003) has, therefore, argued that this last policy phase may herald the end of political concern with regard to alleviating the existing social and economic gap between the Mezzogiorno and the rest of Italy. If this was true, the rate of land abandonment and depopulation of rural areas in the Agri basin may accelerate again - with concurrent negative effects on soil protection practices and desertification as discussed above.

Agricultural policy impacts in the Agri basin

There is little doubt that the CAP and agricultural policy in general has been one of the key policy drivers behind desertification processes in the Agri basin, especially as the subsidy regime linked to the CAP has often led to *environmentally unsustainable intensification of agriculture* in environmentally vulnerable farming areas of the Agri basin. Yet, there have been substantial differences with regard to policy impacts on landuse and farm practice changes in different parts of the Agri basin. Overall, our farm survey revealed that a large number of farms in the Agri had changed farming practices over the last 20 years, often as a result of financial incentives offered by productivist CAP policies, but, as Table 5.5 shows, the extent of changes varies considerably between the lower, middle and upper Agri. Thus, the extent of changing farming practice induced by policy was particularly high in irrigated areas of the lower Agri, where 96% of farmers indicated changing some of their farming practices over the past 20 years (including 86% who improved irrigation systems), and changes were also frequent in the erosion-prone upper Agri where 71% of farmers indicated changing some of their farming practices, with most carrying out general innovations and mechanising their farms.

The case of the middle Agri is particularly interesting with regard to CAP policy drivers. Here, 'only' 52% of farmers indicated policy-induced changes to farming practices, suggesting a slightly less pronounced impact of the policy environment on farm intensification in this part of the Agri basin (see also Anania *et al.*, 1992; Eboli, 1992; Mantino and Turri, 1992). Livestock farming remains the main type of agricultural production in the middle Agri, where several geographical, edaphic and infrastructural factors, as well as lack of market opportunities (especially for sheep and goat meat), reduce options available to farmers to change existing farming practices. A further reason for farmers' reluctance in the middle Agri to invest in costly farming change is both the difficulty of obtaining loans from banks and

Table 5.5: Changes in farming practices (%) (Source: Italian farm survey).

Which changes have occurred on your farm?	Lower Agri	Middle Agri	Upper Agri	Total
Changes in farm practice	96	52	71	65
Changes in farm management	40	11	29	21
Changes in type of farming	68	6	35	25
No significant change	4	48	18	33

the length of time farmers have to wait before receiving CAP subsidy payments. The middle Agri, therefore, emerges as a more static area with regard to policy drivers for agricultural change, with less innovative and 'risky' agriculture remaining predominant. Considering that our sample focused on relatively large farms (see Section 5.1), this result is even more significant as these holdings represent farms that would normally be more likely to innovate or change farming practices based on better capitalisation than smaller farms.

As highlighted in previous sections, one of the key results of our survey linked to agricultural policy drivers is undoubtedly the large impacts that the CMO for durum wheat has had for landuse change in the Agri basin and for exacerbating the threat of desertification - partly linked to the predominance of the *productivist policy agenda* in the area as outlined above. Indeed, durum wheat subsidies emerge as a key driver for landuse change and land degradation issues in the Agri. Several stakeholders admitted that the subsidy led to an overuse of land, especially through the ploughing of marginal and fragile areas (badlands) previously covered by maquis vegetation that traditionally acted as a crucial protection of these desertification-prone areas from soil erosion. Farmers argued that cultivation of durum wheat largely takes place in order to 'crop the subsidy', even in areas where it is virtually impossible to successfully cultivate durum wheat (Arcieri et al., 2002). However, durum wheat subsidies can also be seen to have positive effects with regard to land degradation issues as they are a key financial source for keeping farmers on the land. Indeed, in many farming areas of the Agri, the cultivation of durum wheat subsidy represents one of the few reasons to continue with farming activities, and several authors (e.g. Formica, 1975; Rossi-Doria, 2003) argue that durum wheat is the only *profitable* cultivation choice in inner hilly areas of the Agri. As many farm respondents argued, if durum wheat subsidies are substantially reduced or abandoned altogether in the wake of current CAP reforms, many farmers in the Agri will abandon farming altogether. In this context, Conforti (2002) considered different scenarios for subsidy reduction and concluded that any reduction would lead to further social imbalances, especially linked to socio-economic issues of depopulation and social polarisation outlined above.

Thus, in the middle part of the Agri basin, in particular, most of the strategic options regarding farmers' choices have been deeply affected by the CMOs for both durum wheat cultivation and cattle breeding, especially as subsidies regarding farm mechanisation and/or irrigation equipment have been much less important in this area than in the upper and lower Agri. Nonetheless, subsidies for durum wheat cultivation are also important in the upper Agri, where 23% of our respondents saw it as influential for helping farm survival. In the lower Agri, meanwhile, subsidies for mechanisation and irrigation equipment, as well as CAP payments encouraging crop variation, play a much more important role with nearly a quarter of farmers respectively highlighting that these types of subsidies are crucial for their farming activities.

The situation regarding the importance of sheep and goat meat CMOs is entirely different. Our farm survey highlighted that few farmers were well informed about these subsidies, despite of the dominance of livestock farming in the middle and upper Agri (in contrast to findings of our Greek case study area presented in Chapter 7). Our survey also emphasised the almost complete absence of market organisation for sheep and goat meat in the whole of southern Italy. As a result, Agri farmers producing sheep and goat meat sell their products directly to large traders or cooperatives from northern or central Italy who define the trading price without negotiation. In this respect, the sheep and goat meat subsidy is not a crucial financial help, nor is it a specific reason for sheep and goat farmers to stay in farming (in contrast to those affected by durum wheat subsidies). Sheep and goat farmers emphasised during interviews that a key reason for them to continue with farming is that cooperatives defend their interests by making them more competitive on the national market and by helping obtain the maximum prices for their products - a situation similar to that of farmers in the lower Agri with regard to horticultural production. A further key difference to durum wheat (and other cereal) farming is that livestock breeding in the Agri usually constitutes a full-time activity, demanding the presence of the farmer all year long. As Cembalo (1980) has highlighted, this means that the willingness to carry on the activity without a decent income is much lower among these farmers than for other types of farming - with or without subsidies. Overall, livestock farming in the Agri has experienced dramatic changes over the last decades. In the 1960s, while the rest of Italy experienced a reduction of livestock, the trend in southern Italy has been positive (Cembalo, 1980). Above-mentioned CASMEZ funds greatly aided in the transformation of small farms into large milk or beef livestock units based on the model operating in northern Italy, although this intensive model did not work as well in the context of the Agri basin as it implied a consistent need for water to irrigate alfalfa and maize fodder crops (Arcieri et al., 2002).

Agri-environmental policies form the last cluster of policies with important repercussions for landuse change and desertification in the Agri basin, and can be described as relatively successful with regard to desertification mitigation, albeit with large regional disparities within the region. The agri-environmental programme of Basilicata, approved in 1994, has particularly provided financial aid for integrated farming practices, extensification, and stocking reductions - the latter two particularly important for desertification mitigation. The most recent data suggests a relatively successful uptake for integrated agriculture (95,600 ha enrolled), followed by 20-year set-aside (22,600 ha) and by schemes promoting extensive agriculture (12,800 ha) (Arcieri et al., 2002). More than two thirds of our respondents from the lower Agri, for example, received support through agri-environmental Regulation 2078/92 (Ferraretto et al., 2003).

Overall, agri-environmental measures seem to be more successful with regard to uptake in the more dynamic farming areas of the Agri, especially in the lower Agri, while AEP has only had limited success in the middle and upper Agri. Table 5.6 shows the gap between farmers' knowledge of agri-environmental incentives between the lower Agri and the rest of the Agri basin. Thus, 72% of farmers in the lower Agri had started with application procedures for agri-environmental scheme participation, with a focus on schemes supporting integrated farming, while in the middle Agri fewer farmers decided to start application procedures (41%), with the main reasons being pessimism about the chance to have their applications accepted, and with 62% of respondents arguing that the bureaucracy associated with scheme participation would be too demanding on their time. However, as many studies have shown (e.g. Buller et al., 2000; Wilson and Hart, 2000), agri-environmental scheme participation does not guarantee successful implementation of environmentally sustainable agricultural practices. In the Agri, as in many other European farming regions, a key concern is the lack

Table 5.6: Knowledge about Regulations 2078/92 and 2080/92 (%) (Source: Italian farm survey).

Are you aware of EU Regulations 2078 and 2080[1]?	Lower Agri	Middle Agri	Upper Agri	Total
Yes, I have started the application procedure	72	41	6	42
Yes, I envisage participating in the near future	16	3	6	7
Yes, I am aware	8	13	59	19
Yes, but I do not intend to participate	4	41	6	26
No, I do not know what they area	0	3	24	6

[1]Regulation 2078 = Agri-environment regulation; Regulation 2080 = Forestry regulation promoting the planting of forest (also on farmland)

of adequate monitoring of farmers' actions and inactions. Respondents from the local farmers' union admitted that some Agri farmers did not adhere to agri-environmental prescriptions, and that, as a result, better compliance monitoring is needed to make these policies more effective (Arcieri *et al.*, 2002).

Finally, there have also been severe concerns about the impact of set-aside policies in the Agri basin. The uptake of set-aside has been relatively high in the Agri basin, especially in the middle Agri where 20% of farmers are participating. The aim of set-aside in the area has been to reduce soil erosion based on reduction of arable cultivation, with potentially important benefits for desertification mitigation. Yet, as some stakeholders have stressed, the above-mentioned subsidies for durum wheat cultivation compete with the set-aside measure and reduce its relative profitability, leading to a conflict of interest for many local farmers. Most worryingly, however, is the fact that some local stakeholders claim that set-aside may have exacerbated desertification instead of mitigating it, as protective scrubland areas have been ploughed by farmers in order to be subsequently eligible for 'setting-aside' these areas and claiming the subsidies! This latter point highlights that statistics on successful uptake of agri-environmental measures need to be treated with extreme caution, as destructive farming activities preceding the decision to take part in a scheme (in this case destruction of protective vegetation cover) may have initially exacerbated desertification risks in some parts of the Agri basin.

5.5 Conclusions

This chapter has highlighted that policy drivers for desertification in the Italian case study area have been different from those of the Spanish case study discussed in Chapter 4. Thus, in the Agri basin two major policy drivers have exacerbated desertification processes. First, although partly successful, social and economic policies have not always managed to halt the trend of depopulation in many rural areas of the Basilicata region. The fact that farmers abandon their farms for better opportunities in urban areas has meant that traditional, and often environmentally sustainable, farm management practices have been abandoned, thereby allowing desertification processes to accelerate in already erosion-prone central parts of the Agri basin. Second, the CAP emerges as a key driver for intensification of arable production in the Agri basin, especially through subsidies linked to the durum wheat CMO. This has encouraged many farmers to cultivate wheat more intensively than before and/or to plough up new, and often environmentally vulnerable, areas for expansion of highly profitable durum wheat cultivation, with detrimental repercussions for desertification in the area.

Both our stakeholder and farmer interviews in the Agri basin highlighted that social and economic policies in southern Italy have helped this area to keep pace with the richer northern and central Italian regions during the last decades of the 20th century. However, these policies had different effects in different parts of the Agri basin. Inner areas, for example, continue to lack public services necessary to enable people to live in rural areas with a reasonable quality of life. Large infrastructure projects have partly backfired, in that they have helped people leave their farms for the cities, rather than helping them reach an acceptable quality of life in their own rural areas. This can be largely blamed on the lack of secondary infrastructure (i.e. secondary roads, public transport and other public services), which are fundamental to ensure a decent quality of life in still relatively impoverished rural areas of southern Italy. The failure of social and economic policies in the recent past has led to migration phenomena that have deeply influenced traditional landscape management and conservation in inner areas of the Agri. These trends have been exacerbated by the growth of tourism in coastal areas of the Basilicata region that acts as a further magnet for poor rural populations of the inner Agri, as well as worsening land degradation processes in coastal areas (especially with regard to unsustainable increases in water consumption).

Throughout this chapter, we have emphasised that agricultural policies have been a key driving force for landuse change in the Agri, although their effects have varied depending on farm structural factors such as farm size and type of farming. Thus, large holdings in inner areas are largely driven by CAP subsidies in their landuse decision-making, while smaller farms have been largely bypassed by policy incentives offered through the CAP. This sends a clear message to future policy-makers who will have to ensure that policies are put in place that, on the one hand, ensure survival of farms that still farm in traditional and environmentally sustainable ways and, on the other hand, provide some penalties (rather than rewards) for indiscriminate intensification of arable farming - the latter leading to further depletion of soils and accelerated desertification processes in areas already vulnerable to severe land degradation processes.

At a broader level, our results have shown that national and regional policies for natural resources management have not yet substantially alleviated desertification in the Agri, especially as they operate under a relatively unclear framework of both decision-making and implementation powers with a multiplicity of uncoordinated authorities. Further, coordinated policy activities to tackle desertification have only recently taken place. In this context, it is important to note that in June 1997 the Italian government finally ratified the UNCCD framework (see Chapter 2) through 'National Law 170'. Subsequently, a committee to implement the UNCCD - the *Italian Committee to Combat Desertification* (ICCD) - was established in September 1997 with the specific task of coordinating national and local activities and preparing and implementing the Italian NAP. The ICCD has already carried out many activities and initiatives, including workshops and international meetings, and initial initiatives appear promising for the future with regard to improved desertification mitigation. In particular, the ICCD has established a *National Observatory on Desertification* at Asinara, as well as a *Centre on Traditional Knowledge* in Matera - the latter emphasising the shift that is taking place in Italian governance structures that now place more emphasis on using local traditional knowledge for helping to combat desertification.

At the end of 1999, the NAP was approved with four key foci for combating desertification including soil protection, sustainable management of water resources, reduction of environmentally unsustainable impacts of agricultural activities, and land restoration. Further, the NAP has assigned a crucial role to individual administrative regions and basin authorities in designing specific programmes aimed at combating drought and desertification at regional level. Yet, only the future will tell whether these new policy developments will act as a crucial counter-balancing force to 20th century policies that, as this chapter has shown, have tended to exacerbate rather than alleviate desertification in Italy.

6. Desertification and policies in Portugal: landuse changes and pressures on local biodiversity

M. Vieira and P. Eden

In this chapter we will analyse the effects of past policies on desertification in the region of the Alentejo in Portugal. Section 6.1 will briefly describe the Alentejo and will give an overview of methodologies used during our research (see also Chapter 3 for overall methodological approaches used). Section 6.2 will give a brief overview of key policy phases influencing landuse change in Portugal and the Alentejo. Structural and environmental characteristics of the Alentejo will be analysed in Section 6.3 as a basis for understanding problems associated with desertification mitigation in the area. Specific focus will be placed here on socio-economic and geographical features of the Alentejo and on discussing the current status of protective vegetation cover. The bulk of this chapter will investigate the interlinkages between desertification issues and policies in the Alentejo in Section 6.4. Here, the focus will be on discussing CAP policies as drivers of change, analysing recent policy drivers of landuse change and desertification, and assessing stakeholder perceptions of desertification. Desertification issues and the recent policy arena will also be discussed by investigating data availability for assessment of policy impacts, by assessing policy design and implementation issues, and by analysing the impact of policies on landuse change in the Alentejo using the specific case of agricultural policies. Concluding remarks will be given in Section 6.5.

6.1 The Alentejo: case study area and methodology

The Portuguese case study area covers the whole of the Alentejo region comprising some 2,700,000 ha (Figure 6.1). Rural landscapes of the Alentejo have been moulded by human activity over many centuries, and the evolution of the main farming systems has depended on local soil and climatic conditions (Albuquerque, 1961; Comissao Nacional do Ambiente, 1982). Environmental change induced by human action has created a cause-effect relationship that has frequently resulted in degradation of soil and water resources, and that has ultimately led to a reduction of farming and livestock activities with a lowering of both productivity and the resilience of local ecosystems (Roxo and Mourao, 1994; Loureiro *et al.*, 1994).

Agricultural systems found in the Alentejo can be divided into four principal types. First, the *montados* of cork and holm oak occupy an area of 1 million ha (36% of total area of the region). This system is based on environmentally sustainable and multifunctional use of soil resources, where open woodlands are used for both pastoral and arable farming (Mendes, 1992). Second, permanent pasture forms an important farming system in the Alentejo, usually with spontaneous grassland and shrubs, covering about 500,000 ha (18% of total area). Arable crops form the third farm management system, mostly comprised of dryland

Figure 6.1: Location of the Alentejo case study area (Source: authors).

cereals of soft wheat, durum wheat and oats, and covering about the same area as permanent pastures (Feio, 1993). The fourth agricultural system includes permanent crops of olive groves and vineyards, occupying about 160,000 ha (5% of the Alentejo). These four agricultural systems cover around 80% of the area, highlighting the predominance of agriculture in the Alentejo over other landuses. This means that farming practices shaping these management systems are pivotal for protection and degradation of soils and plant cover.

The best quality soils on flat or undulating land are generally used for cereals, sunflowers, irrigated crops, forage crops, livestock grazing, olive groves and vineyards, while medium quality soils on flat or undulating land are characterised by cultivation of cereals, forage crops, pastures, cork and holm oak montados, livestock grazing, olive groves and vineyards. Poor quality soils and other soil types, meanwhile, are usually found on steeper slopes often prone to erosion or on soils affected by water logging. These latter soils usually support pastures, montados, areas for livestock grazing, forestry, and scrublands, and are, arguably, the most vulnerable with regard to landuse change drivers and desertification processes.

These agricultural systems and their varied susceptibility to erosion and desertification already highlight that the Alentejo forms and ideal case study for the analysis of policy effects on desertification processes in Portugal. Overall, there were five main reasons for selecting the Alentejo as a case study area for our research in the MEDACTION project. First, the Alentejo has a relatively homogeneous landscape that is representative of the Portuguese Mediterranean climate and vegetation. This will allow us to generalise from our results on policy effects beyond the case study region itself. Second, it has the climatic conditions that make it potentially very vulnerable to desertification phenomena. Third, cereal and livestock farming and silviculture are carried out over the whole area of the Alentejo, and all three farming systems can simultaneously contribute towards exacerbating or mitigating desertification. The Alentejo is, therefore, an ideal area to study policy drivers with regard to factors that enhance, as well as alleviate, desertification processes. Fourth, large changes with regard to both social and agricultural policies have occurred, leading to rapid and profound alterations in the management of crop cover, vegetation and soils, with resulting effects on desertification. Fifth, application of results from the Alentejo could be of great benefit towards understanding desertification in Portugal as a whole, and in the adoption of general measures to help mitigate the degradation of soils and local plant cover. This is facilitated by the fact that most land management processes in the Alentejo are linked to agricultural activities that contribute either towards the protection or degradation of soils and vegetation cover. Indeed, there is no other Portuguese region where the agricultural systems have such an important influence on the management of the territory.

In order to collect data on policy drivers behind desertification, we interviewed 45 stakeholders and surveyed 121 farmers in the Alentejo. Interviewed stakeholders were representatives of the general state administration and regional administration (n=13); higher education colleges (5); farmers' organisations (19); environmental NGOs (2); and town or city mayors (6). The selection of farms for the survey was made according to two criteria (see also Chapter 3). First, half of the farms should be situated in areas of high susceptibility to desertification (see Figure 6.2 below) so that differences in types of farm management could be identified and compared with other farms in areas of medium susceptibility to desertification. Second, while taking into account that many of the holdings in the Alentejo are mixed farms that may include montados, cereals, livestock and olives, for example, our farm selection ensured that each farm was representative of at least one of the four main types of farming systems outlined above and considered important for desertification processes.

6.2 Key policy phases influencing landuse change in Portugal and the Alentejo

Having briefly described the region of the Alentejo and the methodologies used in our study in the previous section, in this section we wish to briefly describe the key policy phases that have influenced landuse change in the Alentejo to set the scene for the remaining analysis in this chapter (a more detailed policy analysis will be conducted in Section 6.4). In line with approaches used in the other more recent EU accession countries of Spain (see Chapter 4) and Greece (see Chapter 7), our timeframe for analysis broadly covered the time between 1980 and 2002 (the latter was the cut-off date for our collection of data), and most of our analysis will focus on this recent time period. However, as the historical process of human occupation and landuse of the Alentejo region has determined most of the current human causes behind desertification processes, it is worth briefly visiting policy developments that have occurred over the past 70 years.

Two important policy phases can be identified that have either contributed to substantial landuse change in Portugal or that have influenced policy directions for the 1980s and 1990s. First, the 'wheat campaign' is a good starting point for this discussion. Launched in the early 1930s as a political attempt to make Portugal self-sufficient in cereals, with the state paying high guaranteed prices for grain, this early policy resulted in all types of soils being sown in cereals while, at the same time, often destroying traditional management systems and causing irreparable erosion problems. This policy, implemented through the regional services of the *Ministry of Agriculture*, was the follow-up of similar but less important initiatives of the 'first republic' before 1930, and resulted in central government creating a state monopoly for the cereal market. In Section 6.4, we will see how this policy 'pathway' has continued to have severe repercussions for how agricultural land is managed in Portugal up to the present day.

The second major policy phase occurred during the agrarian reform which was a result of the 1974 revolution. This was a brief period during which agricultural investment was minimal and land ownership uncertain. This political crisis caused further destruction of traditional management systems and poor soil management, resulting in increased soil erosion. This 'reform' was decided at central level by political forces involved in the struggle for the control of the state, and implemented by local syndicates that excluded farming communities from political decision-making processes. Although during this period the main types of landuse continued as in previous years without substantial shifts in cultivation patterns, this policy phase also had important repercussions for policies implemented in the 1980s and beyond.

In our case study of the Alentejo, we will focus largely on policy impacts on desertification since 1980, and four key policy periods can be identified that have had substantial impacts on landuse change. The first policy phase of pre-EU accession lasted from about 1980 to 1985. This period was still heavily influenced by the above-mentioned policy changes emanating from the 1974 Portuguese revolution and, as a result, there was a lack of capital and confidence in agricultural investment (bank borrowing interest rates, for example, stood at 30%). Portuguese agriculture was still backward and labour-intensive, with a quarter of the total Portuguese labour force working in the primary sector. Protectionist policies adopted by the Portuguese state meant that high prices paid by the state for cereals made even a crop of wheat yielding only one tonne per hectare profitable. This caused disastrous erosion problems because even the thinnest soils were ploughed up. As Section 6.3 will highlight, clearing the montados of holm oak forest was a frequent phenomenon in this period to enable mechanised cereal farming, exacerbating a trend that had already become apparent in previous decades (Roxo and Mourao, 1994).

The second recent key policy period affecting landuse in the Alentejo lasted from 1986 to 1992, and was characterised by Portuguese accession to the EU and inclusion into the CAP. This period was characterised by EU policies for capital aid for farm modernisation, mechanisation and intensification. As a result, the number of tractors, for example, more than doubled during this period. These policies had both positive and negative impacts on desertification, since, on the one hand, they represented an important support for the revitalisation of Portuguese farms, but, on the other hand, also promoted more intensive use of soils and associated erosion problems. In particular, and as Section 6.4 will analyse in more detail, EU subsidies for cereals and livestock led to intensification of cereal farming and concurrent abandonment of traditional crop rotation practices, resulting in an increase in cropping areas and decrease in fallow pasture areas (Diniz, 1997). This period also witnessed a substantial increase in fertiliser, herbicide and pesticide use, and the area of catch-crops, such as sunflower, was also expanded considerably.

The third policy period occurred from 1993 to 1999 and was largely based on CAP reforms (especially in the aftermath of the 1992 MacSharry reforms) and associated extensification policies. This period led to both extensification of farming in the Alentejo as a result of the decrease of direct subsidies for crops and to the introduction of agri-environmental measures after 1994 (Fino, 1993). As a result, the area in cereals was reduced, areas of fallow pastureland increased, and animal stocking rates per hectare decreased (Buller *et al.*, 2000). This period also witnessed some positive changes in soil management, with less use of the mouldboard plough and more common use of chisel-type ploughs or heavy discs, which, overall, led to some reduction of soil run-off. At the same time, however, better soils saw an increase in irrigated crops such as maize, tomatoes, melons and sunflowers. As Section 6.3 will discuss, through the introduction of EU Regulation 2080/92 to afforest farmland, the planting of new trees had, in general, a positive effect on desertification and soil erosion.

The final policy phase (since 2000) is linked to CAP reforms through Agenda 2000. Due to further cuts in the subsidy for soft wheat, many cereal farms changed from soft wheat to durum wheat, further encouraged by the higher subsidy being offered for this crop (see also Chapter 5 on the Agri basin case study area in Italy where similar patterns have occurred). This particularly led to maintenance of high levels of intensive soil use on cereal farms (Peixoto, 1998). Based on new water policies, the Portuguese government also promoted the increase of irrigated areas, with investments such as the Alqueva dam scheme on the river Guadiana in the Alentejo (completed in 2002), which created the largest artificial lake in the EU. Once fully operating, the reservoir will irrigate another 110,000 ha of intensive crops, although it is not clear at the time of writing (late 2004) whether this reservoir will be viable in the future, as the availability of production quotas could be a limiting factor on water demand.

6.3 Structural and environmental characteristics of the Alentejo

The impacts of policies on desertification in the Alentejo can only be fully understood if we take into account structural and environmental characteristics of the Alentejo. Thus, before analysing the interlinkages between desertification issues and the policy environment in more detail in Section 6.4, this section will briefly analyse socio-economic and geographical features of the area (Section 6.3.1) and the current status of protective vegetation cover in the Alentejo and the role of forest fires (Section 6.3.2).

6.3.1 Socio-economic and geographical features

The Alentejo has a small percentage of the national population (5.2%) in a relatively large area, resulting in a low population density of only 19.2 people/km^2. This makes the Alentejo the least populated region of the country. As in the Italian case study analysed in Chapter 5, there has been a steady decline of population overall combined with a relative increase in the elderly population over the past 40 years, due to the combination of an exodus of younger people to more economically developed regions, the crisis in local employment essentially dependent on the primary sector, and the negative balance between birth and death rates.

The geography of the Alentejo is relatively unique and explains to some extent why the area is prone to desertification problems. Consisting of a raised plain, the Alentejo region extends from the river Tagus in the north to the Algarve hills to the south (see Figure 6.1 above). Most of the rainfall comes from the Atlantic Ocean in the west, and to the east the area is bounded by the borders of the Spanish provinces of Extremadura and Andalucia. With more than 2,700,000 ha of surface area, the Alentejo is the most distinctive Portuguese region in terms of landscape, both with regard to physical aspect and natural vegetation. This distinctiveness is also true for the different natural or anthropogenic ecosystems that exist in the Alentejo, including the unique agri-sylvi-pastoral system of the montados shaped by human management over time (see below), open cereal fields, olive groves, steep river valleys, and scrublands - all of which play an important role for maintenance of the relatively high biodiversity that can be found in the Alentejo.

The climate of the Alentejo is temperate, typically Mediterranean and continental with hot and dry summers. There are local climatic differences depending on proximity to the coast, height above sea level and exposure to Atlantic winds. The Alentejo coastal region has a higher rainfall and less thermal variation and less extreme maximum and minimum temperatures than interior areas, linked to the influence of winds from the south-west which carry humid air masses. Average sunlight hours are high, with above 3,000 hours annually. The rainfall pattern is notoriously irregular and often falls in torrential showers, one of the main reasons for the high soil erosion rates that occur in the region (Geeson *et al.*, 2002). The average altitude is generally around 200 metres with a few hilly formations such as Marvão or Alandroal.

Soil types in the Alentejo are generally shallow and acidic, with poor drainage and low quantities of organic material (less than 1%). They originate from schist, sands, limestone and granites. Tables 6.1 and 6.2 show the different soil types found in the Alentejo and the distribution of soils of different quality. Only about one third of the Alentejo soils are of sufficient quality to be suitable for cereal growing. The principal limitations for soil use in the Alentejo are very high erosion risk, summer water deficit, low fertility, and rocky outcrops. The region is also vulnerable to repeated drought periods, which have impacts on agricultural production, the well-being of local ecosystems and the quantity and quality of public water supplies. The vulnerability of the local agri-systems to drought is a factor in the stagnation of the rural economy and depopulation processes, and acts as a pressure factor to force changes in agricultural systems in a constant battle for economic survival.

Table 6.1: Main types of soils in the Alentejo (Source: INE, 1999).

Soil type	Percentage of total area
Loams	37%
Lithosoils and Podsoils	30%
Litolicos	16%
Clays	12%

Table 6.2: Soil classification in the Alentejo (INE, 1999).

Soil classification (A = best quality; E = poorest quality)	
A	4.5%
B	10.1%
C	18.1%
D	25.6%
E	41.7%

6.3.2 The current status of protective vegetation cover in the Alentejo

A healthy vegetation cover is a fundamental instrument in defending the soil against erosion, as well as improving fertility and producing food. This is all the more important in arid or semi-arid regions where the adaptation of the vegetation to edaphic-climatic conditions is fundamental (Geeson *et al.*, 2002). Plants with a long life-cycle such as trees can provide an important source of food for herbivorous animals, have an important additional role in the protection of the soil surface, provide shade, and form an important ecosystem for wildlife while, simultaneously, helping to regulate the hydrological cycle, climatic balance, and habitat diversity. The condition of the cross-section of trees surviving in the region can be seen as an important indicator for desertification phenomena affecting the Alentejo (see below).

During the last centuries agricultural landuse in the Alentejo has brought about the destruction of large parts of the natural vegetation, in particular oak woodland and riparian corridors. Species such as holm oak (*Quercus rotundifolia*), cork oak (*Quercus suber*), alder (*Alnus glutinosa*), ash (*Fraxinus angustifolia*) and elm (*Ulmus sp*) were all severely affected by human clearance, resulting in a dramatic reduction in tree densities. Today, Alentejo woodlands cover only about 30% of their original area.

Over centuries, the Alentejo *montados* have been of particular importance with regard to land degradation alleviation. In this agri-sylvi-pastoral system, traditional land management aims to maximise plant growth, while at the same time protecting future production potential.

As a result, the montados are extraordinarily rich in both fauna and flora, and they constitute one of the most valuable habitats in Europe as well as a unique landscape feature. Yet, our survey has shown that almost all montado of cork and holm oak are located on soils of medium or poor quality that generally have slopes that would be liable to erosion if they were not protected by trees. This means that farmers need to find and maintain the balance between woodland and pastures, assuring conditions for the maintenance and regeneration of the system in a sustainable way. As we will discuss in more detail in Section 6.4, this fragile and often unstable balance is frequently upset by market forces, agricultural policy, and by incentives that coerce the farmer into a management direction which produces maximum short-term financial returns, but without consideration of future survival of this environmentally sustainable landuse system. In this context, it is easy to see the importance that agricultural policies have had in land management. As Section 6.4 will discuss, cereal production incentives have, over the years, led to an intensification of landuse of the montados, with frequent and deep soil tillage resulting in an inevitable degradation of the soils in forest areas. Excessive pruning, or the extraction of trees to produce charcoal, also has significantly contributed to degradation of the montado system. Portugal's entry into the EU accentuated many of these processes, especially as EU policies tended to separate woodland, livestock and arable activities where, in the case of the montados, they have coexisted for centuries. In the Alentejo, it is evident that policy compartmentalisation has, therefore, also led to a compartmentalisation of originally sustainable land management systems into unsustainable sub-systems.

The main problems that affect the survival of the Alentejo montados stem from the imbalance brought on by intensification, followed by abandonment of one or more of the components of this system. This is demonstrated by the lack of regeneration within the woodland montados, and by a lowering of both fertility and the capacity of the soil to regenerate, with inevitable loss of biodiversity and habitat mosaics. Ageing trees, decreasing tree density, and a scarcity of young trees, are common symptoms of degradation in the montados. This situation is the result of incorrect management of the land through both arable and livestock activities. Intensification of crop rotations or overstocking also prevents natural regeneration of woodland. Beyond affecting natural regeneration, these practices can also affect the health of individual trees in montados through deep ploughing and the lowering of soil fertility. The consequent weakening of the trees makes them more susceptible to insects and disease, resulting in increased degradation of the woodland. Lowering of the diversity of flora, as well as eroded soils with low levels of fertility, are also the result of intensification of crop rotation and overgrazing. As a result, pastures and woodland lose biodiversity, soils are exposed to erosion, and the less diverse flora in natural pastures leads to an explosive expansion of uniform scrub associations.

As a result of these processes, many farmers indicated in our farm survey that they were generally worried about the high levels of tree mortality in the Alentejo, mostly in oak montados and in olive groves (over the past 3 to 5 years in particular). Farmers saw this as alarming, and argued that, as highlighted above, this may be due to various factors including disease and pests, as well as climate change and deep cultivation methods that damage root systems. Table 6.3 shows rates of tree mortality in the north and south Alentejo, and between montados and olive groves. The data suggests that 77% of montado farms and 61%

Table 6.3: Tree mortality on Alentejo farms (Source: Portuguese farm survey).

Figures represent numbers of farms	Montados (cork and holm oaks)		Olive groves	
	Farmers interviewed	Occurrence of significant tree mortality	Farmers interviewed	Occurrence of significant tree mortality
North Alentejo	6	5	10	8
South Alentejo	25	19	8	3
Total	31	24 (77%)	18	11 (61%)

of olive farmers have problems with tree mortality, with major repercussions for soil protection. In particular the 2002/2003 winter was severe in the Alentejo, with several bad frosts coupled with high rainfall. This might have accentuated tree mortality, especially in olive groves as olive trees are particularly sensitive to waterlogged soils.

Overall, these processes increase the risk of biophysical desertification of the Alentejo, which has the highest levels of susceptibility to desertification as shown in Figure 6.2. This serious situation is aggravated by the occurrence of periods of torrential rain and drought causing

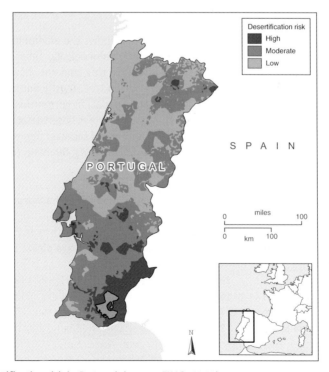

Figure 6.2: Desertification risk in Portugal (source: INAG, 2000).

the greatest damages to farming activities and ecological values of the area. In the Alentejo, the most dangerous periods for erosion on poorer soils and slopes occur every 5 to 10 years when scrub clearance is practiced to control accumulating vegetation that acts as a potential fire hazard. This situation can be exacerbated by the fact that at the time of autumn sowing on poorer soils and on steeper land with a forage crop (e.g. oats, triticale, lupins or permanent clover pasture), the land is susceptible to erosion from torrential autumn rains until the seeds have germinated and have created a binding green cover. Key to environmentally sustainable farming in the Alentejo is, therefore, finding the right balance and timing of specific farm management activities - a balance that has often been struck using traditional farming methods, but that is increasingly being disturbed by policy-driven profit-maximisation cultivation of unsuitable crops (see below).

As can be seen from Table 6.4, cork and holm oak are the dominant species with a wide distribution in the Alentejo, and, as mentioned above, these are the key species that form the agri-sylvi-pastoral system of the montados. The distribution of other forest species, especially eucalyptus, stone pine and maritime pine is generally limited to areas of specific climatic characteristics, depending on altitude and/or closeness to the sea. With regard to our analysis on policy effects (see below), it is, therefore, important to give special attention to policies that influence the regeneration or degradation of stands of cork or holm oak, these being the most important species for the control of desertification. Table 6.5 shows the changes in the area covered by these species between 1980 and 2000, and the data suggests that there is a large increase in forest area of almost 131,000 ha in the Alentejo over this time period, with much planting having occurred on agricultural land.

Yet, despite this positive result overall, Table 6.5 also shows the environmentally potentially disastrous effects of the reduction of area covered by holm oaks by almost 85,000 ha in less than 20 years (see also Table 6.3 above). In addition to this, and not shown in these figures, montados have also witnessed substantial reduction in tree *density* within the stands. This has contributed to the aggravation of biophysical degradation within this type of ecosystem (the Ibero-Mediterranean ecological zone) that has the worst desertification risk in Portugal (see Figure 6.2 above). This represents a significant loss of biodiversity, especially since the montado ecosystem is also protected under the EU *Habitats Directive* (43/92/EEC) and is

Table 6.4: Forest areas by species as a percentage of the total area in each sub-region (Source: National Forests Inventory DGF, 1995).

%	Cork oak	Holm oak	Eucalyptus	Bravo pine	Stone pine	Mixed
North Alentejo	20.8	10.9	6.9	1.9	0.1	7.2
Central Alentejo	15.5	16.2	3.4	0.3	0.4	7.1
South Alentejo	4.2	17.4	1.6	0.1	1.1	3
Coastal Alentejo	24.8	2.6	8.9	5	4.3	10.5
Alentejo	15	12.8	4.7	1.5	1.3	6.5

Table 6.5: Area covered by different tree species in the Alentejo (Source: National Forests Inventory DGF, 1995).

Forest types (by dominant species) (ha)	Alentejo 1980	Alentejo 2000	Difference 1980-2000	
			ha	%
Cork oak	419,500	483,918	64,418	13
Holm oak	482,300	397,811	-84,489	-21
Eucalyptus	41,000	130,520	89,520	69
Maritime pine	31,200	59,479	28,279	48
Stone pine	19,900	52,875	32,975	62

home to many threatened species of flora and fauna. Finally, and arguably most significantly for desertification processes in the Alentejo, there is evidence of the beginning of the end of montados as a major agri-sylvi-pastoral management system, with considerable economic and social implications for the region. As will be discussed in more detail below, the variations in area coverage of different forest types in the Alentejo are, above all, a result of the application of the first and second Community Aid Programmes through the policy on 'forestry measures for agriculture' (EU Regulation 2080/92) and the programme for improving efficiency of agricultural structures (Directive 797/85) (see also Table 2.1 in Chapter 2). As highlighted above, our analysis of policy effects on desertification should, therefore, also pay particular attention to these policies.

When discussing environmental characteristics of the Alentejo, the role of *fires* in shaping the environment warrants specific attention, especially as destructive fires have the potential to destroy the above-mentioned vegetation associations that protect the soil from desertification processes. Forest fires in the Alentejo are more common in stands of eucalyptus and pine, while the occurrence of fires in the montados is low. Native trees, such as holm and cork oaks in the montados, are generally better adapted to fires than monocultural forest stands or introduced species such as eucalyptus, and fire has been an important factor in the evolution of natural plant associations in Mediterranean ecosystems. However, in montado areas where scrub control has been abandoned, the risk of fire increases substantially, and this is usually accompanied by other negative consequences such as destruction of trees, decrease of biodiversity and increased risk of erosion. In particular, the consequences of fires in the Alentejo in stands of cork and holm oak are extremely important. These trees have extremely long growth cycles to maturity, and it is impossible to replace these ecosystems within a human generation. One careless fire may, therefore, destroy plant associations that took previous generations centuries of careful agri-sylvi-pastoral management to create. In the montados, the prevention of fires, or the lowering of fire risk, is largely dependent on good management with grazing cattle and sheep. Here, the way the CAP influences livestock management in a given area is a key factor, as stocking levels or types of animals used (increasingly cattle rather than sheep) are important for the natural regeneration of trees and, to some extent, also influence the vulnerability of certain vegetation associations to fire risk.

6.4 Desertification issues and policies in the Alentejo

The discussion of structural and environmental characteristics of the Alentejo in Section 6.3 allows us to better understand the importance of policy drivers in exacerbating or alleviating desertification processes in the Alentejo. In this final section, we wish to explore in more detail the *interlinkages* between desertification and policies by investigating CAP policies as drivers of change (Section 6.4.1), the impact of recent policies on desertification processes in the Alentejo (Section 6.4.2), stakeholder perceptions of desertification as a basis for understanding different policy 'agendas' (Section 6.4.3), and by assessing the direct links between policy drivers for landuse change and desertification with a specific focus on policy design and implementation issues and the impacts of agricultural policies (Section 6.4.4).

6.4.1 CAP policies as drivers of change

Due to the vulnerable nature of the environment in the Alentejo highlighted in Section 6.3, it is fundamental to evaluate the contribution of CAP policies for their part played in the deterioration or mitigation of the conditions producing desertification. As Table 6.6 shows, variations in UAA in the Alentejo over the period 1989-1995 consisted of a general reduction of annual crops, essentially cereals. This situation is in no way surprising given overall driving forces that have influenced the Portuguese agricultural system (see Section 6.2). While entry of Portugal to the EEC in 1986, and the consequent implementation of CAP policies, initially encouraged agricultural intensification, the CAP reform in 1992 encouraged extensification. These sudden changes in policy direction caught Alentejo farmers unaware, as they were still in the middle of the process of farm modernisation, and made many of the ongoing farm investments unviable. Nonetheless, extensification incentives are the main reasons for the substantial reduction in annual crops and the simultaneous increase in permanent pastures highlighted in Table 6.6.

Table 6.6: Changes in UAA in the Alentejo 1989-1995 (Source: INE, 1999).

	Agricultural use		Difference 1989-1995	
	1989	1995	Hectares	%
Annual crops	635,171	517,458	- 117,713	- 18.5
Permanent crops	169,061	162,019	- 7,042	- 4.1
Permanent pastures	392,077	534,111	+ 142,034	+ 36.2
Fallow	645,853	586,950	- 58,903	- 9.1
UAA	1842,162	1800,538	- 41,562	- 2.2
UAA under trees	715,829	630,809	- 85,020	- 11.8

After 1995, the failure of the Portuguese administration to provide adequate payments for the Accompanying Measures of the CAP Reform (EU Regulations 2078/92, 2079/92 and 2080/92) at levels of payments equivalent to other member states, led to a reduction in income for Alentejo farmers, with further negative effects on traditional farm management practices and maintenance of protective vegetation cover (Vieira and Eden, 2000). This situation has been exacerbated by present policies under Agenda 2000 and the further opening of agricultural markets. Another aspect that has had a great impact on farms in the region is the continuing fall in the prices for the main agricultural products (cereals, legumes, industrial crops, olive oil, cattle, etc.), and simultaneous falls in subsidies for these products. Developments are well illustrated by the price decrease between 1986 and 1997, during which prices overall fell by an average of 40% and, in the case of cereals and beef, by 79% and 50% respectively. Reductions of this magnitude have lowered agricultural incomes substantially in the Alentejo, which is evident in farm and socio-economic statistics. Thus, price developments have had severe structural and social implications in the farming sector. As Table 6.7 shows, many farms in the Alentejo have disappeared and there has been concentration into fewer farms, resulting in larger average size, as well as abandoned areas - all with important repercussions for desertification.

Table 6.7: Farm structural change in the Alentejo (Source: INE, 1999).

Farms by size	1989		1995		Difference 1989-1995	
	numbers	ha	numbers	ha	numbers	ha
Farms overall	45,234	1,842,162	35,673	1,800,536	- 9,561	- 41,561
0<1 ha	6,147	3,555	3,771	2,271	- 2,376	- 1,284
1<5 ha	18,762	43,922	13,460	32,731	- 5,302	- 11,191
5<20 ha	10,413	104,051	8,618	90,342	- 1,795	- 13,709
20<50 ha	3,873	122,530	3,698	120,264	- 175	- 2,266
>50 ha	6,039	1,568,129	6,126	1,554,928	+ 87	-13,201
Average farm area		40.7		50.4		+ 10.3

6.4.2 Recent policy drivers of landuse change and desertification

In Section 6.2 we briefly outlined four key policy phases that have had an influence on landuse change in the Alentejo. In this section, we wish to broaden out the discussion of policy impacts by looking at the specific interlinkages between the most recent policy changes and desertification issues. As highlighted in Section 6.2, Portugal's accession to the EEC in 1986 has been the most recent and the *main cause* of landuse change in both Portugal and the Alentejo over the last 100 years. As Section 6.4.1 emphasised, the CAP initially created a climate for agricultural investment and stability in the Alentejo, resulting in large-scale acquisition of agricultural machinery, which was usually bigger and heavier than necessary. The CAP maintained the same orientation for the production of cereals in the

Alentejo, encouraging increased use of fertilisers, agri-chemicals and irrigation. This resulted in an intensification of cereal rotation and output, with resultant increases in erosion rates and water and soil pollution. Old olive groves, orchards and many montados were destroyed in order to increase areas for irrigated cereals and other highly subsidised annual crops such as sunflower. As a result, this period was clearly negative from an erosion and desertification point-of-view.

In this context, it is important to highlight that the local rural economy of the Alentejo is based on cereal production systems (see Section 6.3). As a result, existent machinery, production technologies and markets are still almost totally geared towards this crop. At the same time, the traditional livestock sector, which has traditionally complemented cereal farming, has suffered a severe crisis due to the Bovine Spongiform Encephalopathy (BSE) crisis and the concurrent reduction in meat demand and production. Together, these factors are already leading to a reduction in pasture area, thereby reversing trends highlighted in Table 6.6 above. The CAP reforms of 1992 focusing on extensification (implemented in 1994) have also led to landuse changes in the Alentejo. For the first time, cereal production has lost its political position as a priority crop, having been granted a transition period of eight years (1997-2004) of gradually declining subsidy support. During this period, the response of both farmers and the administration has been belligerent, attempting by every possible means to keep this traditional crop that has shaped the region's landscape for centuries.

Possible new orientations of landuse, such as environmentally-friendly farming through agri-environmental measures or the afforestation of farmland, are hampered by both the reluctance of farmers to completely change traditional management systems and by the complexity of the proposed agri-environmental policy package (Vieira and Eden, 2000). Moreover, from an economic point-of-view, these alternatives lead to severe reduction of income for most Alentejo farmers when compared to the previous support system, often resulting in the abandonment of farms. Nevertheless, recent changes to the policy-orientation of the CAP have produced some positive environmental impacts such as the increase of biodiversity, the reduction of erosion and water and soil pollution, as well as extensification of livestock production. At the same time, this has also reduced incentives for farmers to further destroy oak trees on the montados.

However, the socio-economic situation of Alentejo farmers and local communities is negatively affected by the reduction of support subsidies, which have not yet been replaced by the new EU decoupled funding - not dissimilar to the situation in the Agri basin in Italy (see Chapter 5). This means that, in the medium and long term, this economic situation will produce negative environmental impacts because, as mentioned above, a certain level of human management is crucial to maintain and restore the landscape systems and to prevent further encroachment from desertification. Our stakeholder interviews made particularly clear that in Portugal, and particularly in the Alentejo, there are currently insufficient management solutions to allow satisfactory implementation of the new decoupled CAP instruments in helping to control desertification. One of the main difficulties in developing such technical and policy solutions derive from the extremely limited understanding and recognition by the administration, farmers and the general public of desertification problems - an issue we turn to in the next section.

6.4.3 Perceptions of desertification

As in the Spanish (Chapter 4) and Italian (Chapter 5) case study areas, our Portuguese study placed great emphasis on understanding different perceptions of desertification among interviewed stakeholders. Obtaining information on these different perceptions helps us to better understand desertification problems caused by specific policies, and why these problems have, at times, not been efficiently tackled (see also Chapter 9).

Our farm survey showed that local farmers have a good understanding of the notion of 'desertification', since the term is used regularly by local media and politicians. Indeed, 88% of surveyed farmers claimed to know the meaning of the term, while 12% could not explain it. Figure 6.3 shows farmers' perceptions of desertification, highlighting that the most important factors deemed to cause desertification are (in descending order) depopulation (seen as nearly twice as important as the 2nd factor), soil erosion, lack of water, degradation of vegetation, deforestation, droughts, climate change, encroachment of deserts, and fires.

As Chapter 2 outlined, according to the UNCCD (1994) the term 'desertification' refers to climatic and/or anthropogenic impacts on soils, leading to irreversible erosion. Answers by

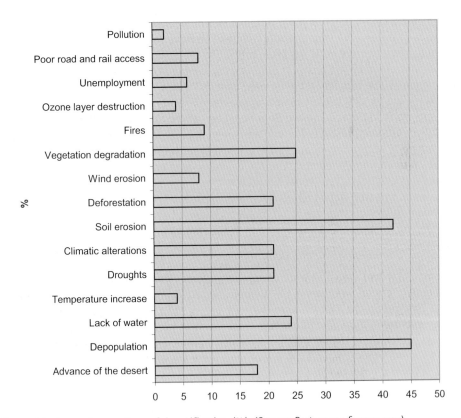

Figure 6.3: Farmers' perceptions of desertification (%) (Source: Portuguese farm survey).

Portuguese farmers with regard to the meaning of 'desertification' included 'soil erosion' and 'climatic factors', suggesting that there is some understanding of 'formal' definitions of desertification. This suggests that farmers have become more aware of the concepts and problems of soil erosion and soil management, largely due to publicity given to the issue of desertification by the Portuguese agri-environmental programme highlighted above. Indeed, several agri-environmental measures have been specifically designed to combat erosion, such as extensification of cereals and livestock, minimal cultivation techniques or direct drilling, as well as restrictions on cultivation on steep slopes and encouraging a winter green cover crop to protect soils from erosion. Unfortunately, and as discussed above, agri-environmental payments are considered as much too low by farmers, resulting in the fact that only few Portuguese farmers have enrolled in these voluntary measures. Nonetheless, ideas and solutions for combating desertification are talked about in the farming community, with or without agri-environmental scheme participation, which suggests that the agri-environmental programme has a positive effect on changing farmers' attitudes (see Wilson, 2004, for discussion of a similar situation with regard to effects of the 'Landcare' programme on Australian farmers' attitudes towards desertification).

As Figure 6.3 (above) shows, and similar to the Italian case study in the Agri basin (see Chapter 5), the most cited cause of desertification in the farm questionnaire was 'depopulation'. This stems from the fact that the Alentejo has suffered a dramatic exodus from the land over the past 10 to 15 years, linked to lack of incentives in agriculture or other rural businesses in the area for young people who tend to move to the industrial zones of Lisbon and Setubal for employment. It has been difficult to get businesses to invest in the Alentejo, as conditions are more attractive in large coastal centres. Consequently, the depopulation phenomenon is the biggest factor in many farmers' minds, as people have 'deserted' the Alentejo region, leaving behind an ageing population. While investments in larger Alentejo towns such as Portalegre, Evora and Beja have been aimed at providing incentives to retain the local population, prospects for rural areas remain bleak. As in the Agri basin in Italy, as long as there is uncertainty about future prices and subsidies, farm labourers will continue to leave - with concurrent repercussions for maintenance of environmentally sustainable land management practices. Yet, some profitable crops are currently expanding, especially vineyards and irrigated crops such as maize and tomatoes. These crops create good seasonal employment, although they are often highly mechanised with mechanical vine pruners to harvesters for grapes and tomatoes, and maize production is now almost completely mechanised.

As Figure 6.3 (above) shows, 'climate change' only scores 7th in farmers' minds as a key driver behind desertification processes in the Alentejo. Some respondents felt that there have been definite changes in the climate over the past years, while others believed that climatic developments are cyclical. One of our farm respondents showed us climate records going back 150 years, which show similar weather patterns and changes as we have today (e.g. in the early 1930s one year recorded 1500 mm of rain, while the average rainfall has always been around 500 mm a year; see also Geeson et al., 2002).

Alentejo farmers were also asked about their perceptions of desertification with regard to Portugal as a whole. Perceptions of areas of Portugal most affected by desertification were

more or less congruent with 'reality' (see Figure 6.2 above), with the majority of respondents arguing that the Alentejo was worst affected (77%), followed by the interior (12%) and south (5%), although there were 6% that felt that north Portugal was the worst affected area. A key question related to farmers' perceptions of desertification on their own farms. As Table 6.8 shows, although most surveyed farmers acknowledged that the Alentejo was the area worst affected by desertification in Portugal, 38% claimed that there were *no* symptoms of desertification on their own farms, while an additional 9% admitted that they did not know whether their own farms were affected by desertification or not. This suggests that nearly half of the Alentejo farmers are not acknowledging that desertification is a possible threat on their farms, thereby contradicting 'official' and 'scientific' estimations of desertification problems in the Alentejo shown in Figure 6.2 above. 'Erosion' and 'water deficiency' were perceived as the key desertification-related problems affecting Alentejo farms, but 62% of the respondents had either taken no measures to mitigate these problems or had not replied to the question on 'corrective measures' in the questionnaire. Some suggested drilling more boreholes as a solution, indicating perceptions linked to conceptualising desertification purely as a water deficit problem (see also Spanish case study in Chapter 4 for similar patterns) and, as Table 6.9 shows, only a small proportion of respondents considered the environmentally most sustainable approach of limiting soil tillage as a solution to combating desertification.

Table 6.8: Farmers' perception of effects of desertification on their farms (Source: Portuguese farm survey).

% answers	Effects of desertification on farms
38	no desertification problems on farm
15	see soil erosion and lack of water as main factors
11	desertification leading to tree deaths
9	don't know
4	desertification leading to land abandonment
3	desertification leading to isolation

Table 6.9: Alentejo farmers' suggestions for combating desertification (Source: Portuguese farm survey).

% answers	Suggestions for combating desertification
33	no reply
32	do nothing
10	minimal or no cultivation
7	drill new boreholes
4	create better conditions for workers

6.4.4 Desertification and the recent policy arena

Data availability for assessment of policy impacts

As Chapter 3 highlighted, research by individual teams involved in the MEDACTION project involved an analysis of published sources that may assist with the process of identifying relevant policies concerning landuse, natural resources management, erosion and desertification. This guided the scope of our analysis of the impact of such policies in the Alentejo region. Unfortunately, there are almost no references dealing directly with the problem of desertification in the Alentejo or Portugal. In this context, it was decided to focus on references that could supply relevant information concerning *landuse change* as a result of policy implementation, given the direct link between landuse change and desertification. Most articles written about desertification in the Alentejo are produced from a purely socio-economic perspective, without outlining and quantifying the relations between the physical deterioration of the natural systems and human response concerning landuse. Selected articles offered quantitative information concerning landuse types and landuse change, insights into the potential and actual results of policies that influence farming activities, forestry and livestock, and the description of management systems for soil and water resources.

Despite availability of some background information, the main problem of analysing the direct effects of policies on erosion and desertification in the Alentejo is the lack of quantitative information and data on indicators that could help to compile a picture of the problem. This is particularly the case for policies between the 1930s and 1970s where most of the useable information is qualitative and relates to the wheat campaign of the early 1930s and the agrarian reform of the 1974 revolution (see Section 6.2). The picture after 1970 is clearer, and there is more information on policies influencing desertification processes in the Alentejo, especially as a result of the tangible impact of these policies on farming activities, forestry, livestock production and water use for irrigation. As highlighted in Sections 6.2 and 6.4.1, particularly significant in this context is the CAP under its different regulations from 1986 onwards (year of Portuguese accession to the EEC). As briefly highlighted above, these policy initiatives were responsible for the destruction of significant areas of oak forest (montado) and for encouraging poor management and excessive use of the soils, resulting in increased erosion.

Other policies considered in our analysis include conservation policies (especially those linked to protected areas), and EU initiatives such as the *Leader* and *Life* programmes. EU directives on biodiversity, birds, nitrates and water are also relevant for our analysis, as is the influence of *National Protected Areas* and areas included in the NATURA 2000 network that also has important effects on landuse. Implementation of most of these conservation policies is relatively recent, but the area protected by these policies (18% of the Alentejo, for example) is significant. In particular, NATURA 2000 legislation may help to control the destruction of a number of important Mediterranean habitats, although it is not yet clear how these measures will be funded.

In our analysis, we also focused on the processes of policy formulation, design and implementation in Portugal, and found that there are no policies specifically designed to combat desertification (as in other case study areas discussed in this book), although some measures have beneficial impacts concerning soil erosion and desertification. These include, in particular, the most recent agri-environmental measures that are only now being implemented (2004), and that are complex with regard to requirements for farmers and implementation and monitoring. Initial results suggest that, so far, uptake by farmers has been relatively low. The new RURIS programme, which aims at afforestation of farmland, is a case in point. So far, this programme offers an inadequate format of support for Alentejo farms, mainly because it was designed without sufficiently taking into consideration the relatively large farm sizes in the region. This situation is not helped by the fact that, as highlighted in Section 6.3, traditional silviculture management techniques usually used for planting new woodlands may lead to soil and biodiversity loss and to increased fire risk.

As part of recent policy developments, the national and regional environmental programmes include measures that may be extremely useful for restoring vegetation around dams and river beds, and for the introduction of management systems for subterranean water systems. Other policies currently being implemented which may also help to control desertification include *hydrological river basin plans* which are in the process of being produced at the time of writing (late 2004), and the *Ecological and Agricultural National Reserve*, a general planning classification of landuse which aims at helping to prevent dramatic changes of management in these areas.

Policy design and implementation

As we have seen in discussions of the Spanish (Chapter 4) and Italian (Chapter 5) case study areas, understanding policy design and implementation processes at the national scale are important for understanding effects of policies and problems with transposition of EU policies to the national scale (see also Hart and Wilson, 1998; Jones and Clark, 2001). In Portugal, the design and implementation of any type of policy is extremely centralised within government bodies, and there is weak or no tradition of consultation. Participation of stakeholders in policy design, such as rural communities, is virtually non-existent. Further, neither NGOs nor the general public are accustomed to making use of existent mechanisms for influencing the policy design processes. It should also be noted that the process of 'spreading the message' about sustainable agricultural practices and practical demonstrations is slow, although some promising developments have recently taken place with regard to information diffusion on the interlinkages between farming and desertification by both the University of Evora and the local *Association for Soil Conservation*.

In recent years, responsibility for implementation of regional policies has gradually shifted towards *Regional Coordination Units* set up by the Ministry of Planning. In the present *Community Support Framework* (CSF), more importance has been given to regional bodies of the different ministries, which are now responsible for the implementation and management of *Regional Operational Programmes* which also include areas of responsibility such as the environment and agriculture. From a desertification perspective, when compared with the previous *Regional Development Programme* that operated from 1994 to 1999, the present CSF includes some, albeit limited, reference to the problem of erosion and desertification and to the relationship between the proposed measures and programmes implemented to solve desertification, while the previous programmes had no references to the problem of desertification at all. However, as in Spain and Italy (see Chapters 4 and 5), as well as in Greece (see Chapter 7), it is fair to say that in Portugal, currently, recommendations of the UNCCD are insufficiently included in different measures and programmes suggested by the CSF to tackle desertification.

The impact of policies on landuse change in the Alentejo: the case of agricultural policies

In order to highlight some of the specific impacts that the policy environment has had on desertification processes, this last section will focus on analysing in more detail the impact of agricultural policies on farm management systems in the Alentejo. What this section will show, in particular, is that policies can have *both* positive and negative effects on desertification processes, and that policies that encourage changes to traditional farm management practices are not necessarily always negative policies with regard to soil protection issues.

According to our stakeholder interviews and the farm level survey, the most important policy drivers for landuse change in the Alentejo have been Directive 797 (see Table 2.1 in Chapter 2) and national policy 1991/2328 (PAMAF[24]). These two policies provided strong incentives for farm-level investment intended to improve farm structures. There was less investment under the 3rd EU support framework since 1999, due to declining confidence in agriculture in general, although the exception in the Alentejo has been strong investment in new vineyards and a continuing expansion of irrigated crops. The impacts on cereals and arable crops caused by Directive 797 and the PAMAF policy have been evident, with clear evidence of environmentally-damaging intensification in the Alentejo. This has been further exacerbated by the national *Programme for Agriculture and Agricultural Development* of the Portuguese *Ministry of Agriculture* and by the cofinancing for soft and durum wheat which, as in the case of Italy (see Chapter 5), have led to an expansion of intensive arable production.

These trends have been, to some extent at least, counterbalanced by set-aside policy, as well as agri-environmental measures for minimal tillage, direct drilling and extensive cereal cultivation, that all appear to have positive effects with regard to desertification mitigation. Cereal production in the Alentejo has been very dependent on area subsidies and national subsidies aimed at alleviating the effects of falling grain prices over the past 10 years. Since area subsidies have declined, and since national subsidies have ceased altogether in 2004, the Alentejo has witnessed dramatic changes in crop management away from soft wheat (traditionally grown in the Alentejo) to durum wheat - a process almost entirely driven by high area subsidy payments for durum wheat. Indeed, the durum wheat subsidy has been so successful that the Portuguese durum wheat quota has for the first time been exceeded in 2003 and 2004, leading to a reduced subsidy payment to individual farmers. As in the Italian case study area (see Chapter 5), the switch towards intensive durum wheat cultivation in

---

[24] Programme of Support for Modernisation in Agriculture and Forestry.

the Alentejo can, therefore, be seen as a good example of subsidy-led farming with potentially disastrous consequences for desertification.

Yet, it should also be noted that current landuse practices are also partly a legacy of past policy regimes. Thus, impacts of policies on Alentejo soils date back to the 1980s and even before, when all soil types were ploughed regularly with dramatic soil erosion consequences on thinner soils and on slopes. This was largely due to national protectionist policies of high prices paid for wheat until the mid 1980s (see Section 6.2). Since the 1990s, however, cereal prices have continued to drop in value and CAP subsidies have also decreased. A consequence of both these cereal income falls and recent policy changes of the 1990s and early 2000s (e.g. agri-environmental measures and extensification) has been that current soil use and management has *much improved* in the Alentejo. Indeed, instead of growing cereals on all soil types, under montados and on steep slopes, cereal farming today is mainly restricted to suitable soils, and consequently there are fewer erosion problems than in the past.

It should also be noted that the impact of increased cereal cultivation in some parts of the Alentejo is not always negative for biodiversity and desertification. Mature cereal crops, leftover post-harvest stubbles, and rotational fallows and set-aside are important for birdlife, in particular for bustards, harriers, partridge, larks, as well as wintering green plover and golden plover. Cereal fields in the Alentejo are also rich in small mammals and insects, and form thus an important hunting area for raptors. The flora diversity generally continues to be rich, with many arable weeds particularly in fallow and set-aside areas. There has also been a noticeable improvement in soil management on cereal farms due to the use of only the most suitable soils for cropping (in contrast to the Italian case study discussed in Chapter 5). This has partly led to reduced erosion risk, as well as a shift towards better tillage techniques such as chisel ploughing and minimal cultivation. Our farm questionnaire survey also showed that there is a good awareness among Alentejo farmers of the problems of erosion (42% of respondents), although the main constraint is having sufficient finance to invest in appropriate machinery, which means that the largest positive improvements with regard to adoption of desertification mitigation techniques are generally found on larger, better capitalised, farms.

With regard to irrigated arable crops (maize, rice, sunflowers, tomatoes, sugar beet), the impact of above-mentioned policies (e.g. Directive 797; the CMO for durum wheat; PAMAF), as well as the *Nitrates Directive* (Directive 676/91/EEC), has been an overall increase in irrigated areas over the past 10 years. As in the Spanish case study described in Chapter 4, this can be a source of both erosion and pollution of subterranean waters and streams, particularly with regard to maize and rice crops which require considerable quantities of water, fertilisers and pesticides. Nonetheless, where arable irrigation takes place on good and relatively flat soils and with controlled water application, the problem of erosion is usually minimal. In addition, some crops like maize and tomatoes can be economically profitable and are a good incentive to keep young farmers on the land and, therefore, aid in the prevention of further depopulation of rural areas.

Olive groves are another important crop influenced by subsidies. The most significant policies in this context are the olive oil CMO and agri-environmental measures for 'integrated production', 'traditional olive groves' and 'organic production'. Olive oil subsidies have had a strong effect on ensuring survival of the most suitable olive groves, and there has also been investment in new and more intensively farmed olive groves, usually with drip irrigation. From a productive and economic point-of-view, these new plantations are more viable than the older traditional plantations, but from an environmental and landscape point-of-view, there is a strong argument to retain traditional plantations. Indeed, traditional olive groves are rich in spontaneous flora and bird life, and are particularly important for owls, stone-curlew, hoopoe, blackcap and thrushes, as well as for small mammals, lizards and snakes. The new plantations lack biodiversity due to intensive management and high chemical input. If olive trees are pulled out and not replanted, the land is usually used for arable crops, which will increase soil erosion problems and reduce biodiversity. However, there are also many traditionally-farmed olive plantations on slopes in the Alentejo which are cultivated too frequently to control weeds, leading to soil erosion. This again shows that we should not always equate desertification issues with replacement of traditional management with new and more intensive agricultural practices, as the former were also often not the types of landuse most suitable for the fragile Alentejo environment. Finally, there is also evidence in our survey that more olive farmers are leaving their plantations in pasture for sheep grazing or using herbicide to control weeds instead of cultivating. These practices are, for example, encouraged under the agri-environmental measure 'traditional olive groves'. The agri-environmental measure 'integrated production' is also proving popular with olive farmers. This helps reduce the use of harmful chemicals and reduces excessive weeding, thereby ensuring a protective vegetation cover against erosion.

Vineyards are another important element of the agricultural landscape in the Alentejo, also with regard to their potentially positive desertification-mitigating characteristics as soil protectors. Policies that have affected grape production are the wine CMO and agri-environmental measures of 'integrated production', 'organic farming', and 'investment policies for planting and machinery'. There has been a rapid expansion of vineyards in the Alentejo (e.g. regions of Portalegre or Estremoz), and vineyards add greatly to landscape diversity, and to desertification alleviation, but are usually relatively biodiversity-poor.

Environmentally sustainable pasture management is another crucial aspect of desertification mitigation in the Alentejo. Pastures often have a high natural value both for flora and fauna, having many species of wild grasses, legumes, flowers and bulbs. The fauna may include steppe birds such as the great and little bustards, sandgrouse, stone curlew, lesser kestrel, as well as game species like red-legged partridge, hares and rabbits. These pasture areas are often on soils susceptible to erosion, where land degradation is controlled by the permanent nature of the herbage and its roots, while the biggest threats to pastures are ploughing or overgrazing. In the Alentejo, permanent pastures are found on land not suitable for arable crops due to poor or shallow soils or steep slopes, in waterlogged areas, or on stony or rocky areas. As highlighted above, pastures are also frequent under the tree canopy of the montados, and sometimes in olive groves.

Policies that affect permanent pastures have been, in particular, subsidies for investment in fences and cattle drinking troughs (Directive 797; PAMAF; AGRO; see above), the LFA Directive, and agri-environmental measures for extensive pastures. In the Alentejo, permanent pastures are grazed by livestock and the stocking density depends on the fertility of the soil, usually with 0.2-0.5 livestock units/ha. These areas are sometimes also used for haymaking where the terrain is suitable. Our farm survey revealed that more farmers than expected apply fertiliser to their permanent pastures (63%), usually a nitrogen and phosphate mix (typical Alentejo soils are low in organic material and phosphate, but high in potash). Recent improvements aim towards better livestock handling, such as fencing and cattle drinking troughs, including small dams which can also be important for attracting wildlife. Some of these areas have been planted recently into new montados under Regulation 2080/92 (see above) and cannot be grazed, although they are cultivated annually for weed control. The main policies affecting numbers of livestock and grazing densities have been LFA subsidies, the cattle, goat and sheep CMOs, and agri-environmental measures for grazing extensification and for keeping rare breeds. Headage subsidies caused an increase in livestock on farms during the 1980s and 1990s, often resulting in overstocking and causing damage to the montados by affecting natural regeneration. However, the CAP reforms of 1992 and 2000 have forced farmers to better control cattle densities in order to be eligible for subsidies, with resultant positive effects for soil erosion.

Finally, and as discussed in various parts of this chapter, the environmentally sustainable agri-sylvi-pastoral system of the *montados* forms a crucial aspect of farm management in the Alentejo. It is, therefore, crucial to understand policy drivers that may affect, and indeed, disrupt, the management of the montados. Policies that have had substantial effects on the management of montados include LFA payments, the agri-environmental measure 'protection of holm oak montados', and the CMOs for cattle, sheep and goats. Until the 1980s, cereals were commonly sown in the montados, which resulted in much visible erosion. Heavy discs were commonly used to control scrub, which destroyed young trees as no care was usually taken by tractor drivers to protect seedlings. Today, nearly all montado farmers are managing their forests to allow natural regeneration. This is crucial, as the montados have been neglected in the past, with a significant lack of young trees, resulting in an ageing forest and with thousands of trees dying each year (see Table 6.5 above). As Table 6.10 shows, farmers today also prune the trees regularly, a traditional process which is important for the health of the trees. Most montados are left in natural pastures and are grazed by livestock, and some farmers sow legumes and apply some fertiliser on the pasture. Table 6.10 also shows that cork oak montados are less exploited for crops and livestock and better managed for regeneration than holm oak montados. This is largely due to the fact that cork continues to be a valuable product, whereas holm oaks 'only' provide acorns, firewood and charcoal.

New montado areas have been planted under Regulation 2080/92 (forestry measures for agriculture; see Table 2.1 in Chapter 2), which are partly aimed at re-establishing large areas of montado that were destroyed through the 'cereal campaign period' from the 1930s until the end of the 1980s (see above). Under Regulation 2080, farmers are paid to plant trees, and they also receive a payment for loss of agricultural income over 20 years during which they are not allowed to use the land for grazing. However, despite initial successes that are partly reverting the trend of holm and cork oak destruction of the last few decades (see Table

Table 6.10: Farm practices on montado farms (Source: Portuguese farm survey).

Montado farm practices	% of montado farmers	
	cork oak montados	holm oak montados
Allow regeneration	77	45
Regular pruning	77	69
Sow legumes	11	31
Sow cereals	11	24
Graze livestock	77	86

6.5 above), the problem remains that costs to replant substantial areas are very high, resulting in the fact that the planting programme is only advancing at slow pace.

6.5 Conclusions

Portugal is a country where several areas are affected by desertification. The south of the country is the most severely affected region due to its geographical characteristics such as climate, relief, geology, soil types and landuse systems, and our case study area of the Alentejo has been selected because it has been particularly prone to desertification processes. This chapter has highlighted that policies have had an important role over time in shaping and influencing landuse change in the Alentejo. However, as was the case with the Spanish and Italian case studies (see Chapters 4 and 5), policy impacts have been varied and complex, with some policies having had clearly 'measurable' and tangible effects, while other impacts have been more subtle and difficult to specify (see also discussion of *implicit* and *explicit* policy impacts in Chapter 2).

Our discussion showed that in the early 1980s Alentejo farmers had no income support except for high guaranteed prices for cereals. After EU accession of Portugal in 1986, CAP subsidies have provided about 60% of the Alentejo farmers' income. This means that over the past two decades, the influence of subsidies has often been a stronger driver for landuse change than the value of the farm produce itself. This has resulted in an unsustainable farming climate in which farmers have been forced to adopt management decisions that were often not the best for the farm or the environment, but that were almost entirely based on economic factors. There is, nonetheless, evidence from the farmer survey and from personal experience of the authors (both of whom are involved in farming) that most of the present policies and measures are beginning to have a positive effect on farmers' decision-making and on good farming practices - 'positive' especially in the context of contributing towards desertification mitigation. Yet, at the time of writing (late 2004), the economic reliance on subsidies remains, and any cuts in farmers' incomes would lead to further land abandonment and creation of even larger properties to be able to survive, which, in turn, would lead to further reductions in rural populations with important repercussions for desertification processes (see also Chapter 5).

Our findings have shown that policy changes that have taken place since 1992, in particular, have been *largely positive* in reducing desertification, especially changes linked to agri-environmental policies. However, lack of uptake by farmers of many positive measures in the Portuguese agri-environmental policy programme has prevented wider positive impacts of these policies to take effect over a larger part of the Portuguese territory. It appears that the main two deterrents for successful implementation of agri-environmental policies have been that payments have been too low to compensate farmers for loss of income based on agricultural and environmental restrictions, as well as the often high cost of investment required to achieve the aims of specific agri-environmental measures (e.g. direct drilling measures for arable farms associated with high costs of machinery). Further, agri-environmental contracts are often so restrictive that farmers have been discouraged from joining existing schemes (Vieira and Eden, 2000).

There are, however, encouraging new policy developments under way. Recently, for example, a new 'good agricultural practices farm code' was introduced, which could have positive effects on desertification in the Alentejo. The code is cross-compliant with both the new five-year agri-environmental contracts and with LFA payments. Under this new code, the maximum stocking rate is two livestock units/ha, and there is an innovative *erosion risk factor* attached to each parcel of land (1 'low' to 5 'high') which is identified on the farmer's annual farm subsidy return. An owner of a 'risk 4' slope, for example, will be prevented from planting annual crops, while new planting of trees or shrubs, or the resowing of pastures, can only be carried out after official permission has been granted. For 'risk 5' slopes, meanwhile, no annual crops can be planted and no pasture sowing can be undertaken, although natural pastures can be improved (but only if this involves no disturbance to the soil) and new trees and shrubs can be planted after permission has been sought.

Despite these positive recent developments, this chapter has also shown that the consequences of desertification are not always easily seen in agricultural areas of the Alentejo, and that it is particularly difficult for farmers to associate specific changes in landuse with the possibility of increasing desertification risks. This situation has been exacerbated by the fact that the notion of 'desertification' is interpreted differently by both the various policy-making stakeholders involved in implementing policies aimed at alleviating desertification risk and by the farmers themselves. This has resulted in serious difficulties in successful implementation of actions and policies to combat and mitigate desertification in the Alentejo.

Nevertheless, the fight against soil erosion and desertification should become an important topic in the Alentejo (and beyond) in the near future, with improved opportunities to implement policies aimed at sustainable farming and at reducing land degradation risks. In this context it is, therefore, fundamental to ensure, first, that an awareness programme for farmers and the general public is implemented, in order to show that urgent steps and measures must be taken to combat the increasing problem of desertification which seriously threatens the basic natural resources such as soil and water in the Alentejo. This programme should particularly target farmers who own and manage fragile and vulnerable agri-sylvi-pastoral systems of the *montados* that form a unique and important habitat, as well as acting as a crucial landuse system for soil protection and desertification mitigation (if properly managed). Second, demonstration projects must be implemented to stress the need to change farming practices and soil and crop management techniques to minimise erosive processes caused by extreme rainfall events. In this context, lessons learned in other severely desertified environments such as Australia - where the 'Landcare programme' offers some interesting grassroots-led solutions towards desertification mitigation at catchment level (see Wilson, 2004, for detail) - may be of particular interest. Finally, the true extent of the state of soil degradation and ecosystem destruction in Portugal should be shown by the media, at the same time offering viable solutions which have been both technically and scientifically tested and that have generated visible results. This should serve as an incentive for both farmers and the public in the implementation of measures and actions to better preserve and conserve natural resources with the aim to alleviate desertification.

7. Desertification and policies in Greece: implementing policy in an environmentally sensitive livestock area

N. Beopoulos and G. Vlahos

This chapter analyses the impact of policies on desertification processes on the island of Lesvos in the Aegean Sea, a part of which formed the case study for our analysis. In line with approaches in other case study areas for the MEDACTION project (see Chapters 3-6), we will place specific emphasis on the policy environment between 1980 and 2000. The introductory section (Section 7.1) will give a brief overview of the geography and economy of Lesvos and will discuss our methodological approach. Section 7.2 will discuss the importance of sheep farming in our case study area and will highlight how sheep farming has shaped the landscape and farming traditions in the area - not always with beneficial impacts with regard to desertification mitigation. Section 7.3 will investigate CAP-related policy drivers and their influence on landuse changes, with a specific focus on four key policy arenas, while Section 7.4 will analyse the impacts of these policies on livestock farming in north-west Lesvos. Section 7.5 will then look specifically at the links between land management practices and desertification processes in the case study area. Section 7.6 will examine how policies and measures favouring the development of sheep farming were implemented in the study area, with a specific focus on the mechanisms that have underlain state intervention for many years and through which policies have been promoted to farmers. Concluding comments are given in Section 7.7.

7.1 Introduction

In this introductory section we will provide some background information on the island of Lesvos where our case study area is located, describe our case study area, and provide some information on our methodology.

7.1.1 The island of Lesvos

The island of Lesvos, the northernmost of the large islands in the Aegean Sea, forms the case study for investigation of effects of policies on desertification in the Greek context (Figure 7.1). Most of the surface of Lesvos is occupied by mountainous areas with the highest peak rising to 968 meters above sea level. Fertile alluvial plains are rare and most valleys are characterised by rivers with seasonal flow patterns. The island is characterised by sharp differentiations both in its natural and agricultural areas. In the north-east there are bare uncultivated expanses of land, most of which are used as pastures. In the south-east, extending towards the centre and the north-east, the island is densely planted with olive groves. The north-west, meanwhile, forms our case study area (see below) and is largely

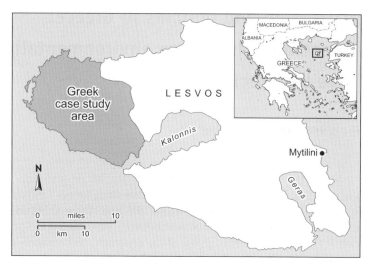

Figure 7.1: The island of Lesvos and location of the Greek case study area (Source: authors).

characterised by extensive pastures for sheep farming. Mountainous areas contain some olive plantations but are largely dominated by pine forests. Occupying around one quarter of the total area of the island, olive groves have slowly expanded, and have become predominant over the last two centuries (Kontis, 1978; Marathianou *et al.*, 2000). Not only the agricultural economy of the island, but also its trade and its industry, are or have been largely based on olive production.

North-western Lesvos will form the basis for our study on interlinkages between policies and desertification, as this area is particularly prone to desertification processes and characterised by only sparse vegetation. The lack of vegetation in this area is a product both of the level of rainfall and of the nature of the terrain and topography, although, as this chapter will analyse in detail, anthropogenic processes have also contributed towards desertification processes in the area. The region has its origins in acid volcanic parent rocks that predominate, and soils are stony, severely to very severely eroded, shallow to very shallow, and slopes are often steep. The combination of these factors makes the north-western parts of Lesvos particularly vulnerable to erosion and desertification (Kosmas *et al.*, 1999a). The rainfall figure for western Lesvos is 45% lower than in eastern areas, with an average annual rainfall of 670 mm in Lesvos as a whole. Vegetation has adapted to survive drought conditions but is also at risk from fires, further heightening the danger of erosion in hilly areas (Kosmas *et al.*, 1999b). In north-western Lesvos, therefore, natural conditions largely determine the way that farmland is utilised and, as a result, pastures and sheep farming are the predominant landuse in the area. At the same time, due to the natural vulnerability of the area to erosion, landuse is a critical factor in exacerbating desertification processes, and sheep pastures are currently the most degraded agricultural areas of the entire island (Arianoutsou-Faraggitaki, 1985).

In the north-west, many national and EU funded research projects have shown the significant danger of desertification (e.g. Margaris, 1987; Yassoglou, 1990; Kosmas *et al.*, 1996, 1999b). It is also argued that the danger of desertification has been exacerbated by the increase in sheep numbers in the area, a development attributed to EU CAP policies which has led to overgrazing, disappearance of vegetation and subsequent erosion (Anthopoulou and Goussios, 1994).

7.1.2 Case study area and methodology

We chose one specific area comprised of eight districts (Eressos, Antissa, Messotopos, Sigri, Chidira, Vatousa, Pterounta and Agra) situated at the end of the island's north-western peninsula as our case study area (see Figure 7.1 above). The key criterion for selection was that, environmentally, this zone is the island's most sensitive area and most threatened by desertification (Kosmas *et al.*, 1999b). For the purpose of selecting the specific case study communities, we superimposed a map of Lesvos showing 'environmentally sensitive areas' prone to desertification - data which was already available within the framework of the MEDALUS EU research programme (see also Chapter 3). Our case study area covered 35,000 ha or 21.5% of the total area of the island. On the basis of three categories of desertification within the ESAs (critical, fragile and under potential risk of erosion), four districts were classified as 'critical', corresponding to the most threatened regions, while the other four districts comprised zones belonging to the other two desertification categories.

Our study began in 2001 with a first research step comprising semi-structured on-site interviews with stakeholders of the region (Beopoulos *et al.*, 2002; see also Chapter 3). Sampling for these interviews was based on a quality supply and demand model of the local production system. The model suggests a system consisting of individual elements where stakeholders 'demand' and 'supply' quality. Thus, 'supply' and 'demand' can be seen as two interdependent poles of the system, although at the local level this system does not operate in isolation but is also exposed to external factors (e.g. policies). In particular, public authorities set up management interventions, which can either interact directly with the environment or influence stakeholders' attitudes towards public interests (through incentives or regulatory actions, for example). The third pole of the quality-system, therefore, consists of administrators (local public authorities and institutions). We interviewed representatives of these three stakeholder arenas, including 15 farmers, 9 local inhabitants (living and/or working in the case study area) and 7 administrative officials and/or extension workers. Following this, we conducted an on-site farm-level survey through a questionnaire in March 2002. Based on the general methodology adopted for Module 2 of the MEDACTION project (see Chapter 3), the sample comprised 100 sheep farmers selected from the eight districts outlined above, and comprising around 7% of the area's 1,491 agricultural holdings (Beopoulos *et al.*, 2003).

It is important to understand human population dynamics in the north-west of Lesvos in order to contextualise some of the findings presented in the following sections. The population of our case study area has been relatively stable demographically since the 1970s (similar to the island as a whole). Statistical data for the north-west suggest that there was a marked fall in population in the 1960s and a more moderate fall in the 1970s, a phenomenon observable in many parts of Greece owing to internal (i.e. within Greece) and external migration. The 1980s and 1990s were a time of relative demographic stability, a pattern different from other LFAs in Greece, particularly those where livestock breeding predominates (NSSG, 1961a; 1971a; 1981a; 1991a; 2001). In this sense, our case study area differs from other case study areas discussed in this book, in particular the Italian and Portuguese study areas that have experienced pronounced population losses (see Chapters 5 and 6).

7.2 The role of sheep farming in the north-west of Lesvos

This section will provide background information on the farming situation in our case study area in the north-west of Lesvos as a basis for understanding the interlinkages between policies and desertification analysed in Sections 7.3-7.6.

Sheep farming is the case study area's only farming activity. It not only contributes to shaping the rural landscape and its identity, but is also of key significance to the regional economy and for keeping people 'on the land'. It is the key economic activity followed by tourism, but despite growth in tourism in recent years income from tourist activity still remains marginal compared to sheep farming. Our questionnaire data highlighted the *traditional character* of sheep breeding, and all farmers questioned indicated that their fathers and grandfathers had also been engaged in the same occupation. It is a finding also reinforced by farmers' answers to the question on why they decided to become livestock farmers, and main reasons cited were 'family tradition' (38%) and, closely related, 'love of this type of work and the related way of life' (34%). However, another key reason was 'absence of alternatives' (38%), an answer related to the region's poor arable potential and lack of other economic opportunities. Crucial for our analysis of policy drivers of landuse in our case study area, only a small proportion (5%) saw CAP subsidies as a key reason for staying in sheep farming.

The significance of sheep farming for the region is evident from the following figures. Greece's EEC accession in 1981 had a decisive influence on its agricultural development, resulting in distinctive pre-entry and post-entry policy periods. Table 7.1 shows that during the pre-entry period, overall livestock numbers (cattle, sheep, goats) doubled between 1961 and 1981 in our case study area (an increase of 109%), with the most significant increases taking place in the 1970s, and with increases in the whole of Lesvos being even higher (134%). By contrast, the post-entry period (data available for 1981-1998) saw a slight reduction in overall livestock numbers in our case study area (-5.8%). These figures are even more revealing in light of the fact that the number of sheep continued to increase (by 19%) in the post-entry period in Lesvos as a whole, albeit not at the same high rate as during the pre-entry period. This shows the important impact that CAP-related sheep subsidies have had for encouraging high sheep numbers on Lesvos. Rises in numbers were more pronounced outside of our case study area, resulting in the fact that there was a slight decline in the study area's contribution to overall sheep farming on the island. Thus, while in 1961 38.5% of the island's sheep were concentrated in the north-west, by 1998 this proportion had fallen to 'only' 26%. Nonetheless, sheep farming continues to be the key agricultural activity in the north-west, comprising about 94% of all livestock, while goats make up about 6% of the animals.

Table 7.1: Sheep numbers in Lesvos and in our case study area (1961-1991) (Source: NSSG, 1961b, 1971b, 1981b, 1991b).

	1961	1971	1981	1991
Lesvos	117,858	167,599	276,240	267,132
Case study area	42,409	56,202	88,732	83,560

Farms in the study area are mainly family farms with a long tradition in sheep farming, and with most depending entirely on the labour of family members. The 'lesbian' (i.e. from Lesvos!) sheep breed predominates, regarded as particularly appropriate for grazing on poor quality pastures. The farm economy is primarily based on the sale of sheep milk for cheese production and, to a lesser extent, on the sale of lamb meat (it is worth noting here that three cheeses are produced on the island that fall into the category of 'protected denomination of origin'). As evident from our farm questionnaire, farmers do not appear to be seeking to maximise milk production, with per capita production remaining low (108 kg/sheep). Sheep are mainly held outdoors and left to forage by themselves. This means that sheepfolds on farms are often makeshift lightweight constructions, with only very rudimentary mechanical equipment used on farms. Given that pasture areas are often of inadequate size in view of the high sheep numbers (see below), dietary requirements of the sheep are supplemented by purchased concentrated animal feed that are usually cheap by-products of cotton and sugar beet processing. Animal feed is, therefore, selected on the basis of low cost and not for considerations of dietary balance or nutrition.

Pasture covers 81% of the territory of the study area (40% for the whole of Lesvos), forest accounts for only 4.8% of the area, while cultivated land covers 8% and is generally found in small valleys adjacent to the coast (NSSG, 1991c). Cultivated land includes mainly olive plantations, some fig trees and some annual crops for livestock fodder. For historical reasons, land ownership patterns in Lesvos differ from those of mainland Greece, with private pastureland comprising 92% of the total, in contrast to only 42% for Greece as a whole (with the remainder belonging either to communities or the state). However, despite this high level of private ownership in the study area, according to data from the questionnaire most farmers rent pasture from private individuals (84%), and for three-quarters of our farm respondents rented pastureland makes up over half of their UAA (Beopoulos et al., 2003). As far as management of pastures is concerned, questionnaire data shows that average grazing intensity is 0.68 LU/ha. If we also take into account that land under olive trees is often used for grazing sheep, the grazing load falls to 0.59 LU/ha.

The use of fire in pasture management in north-west Lesvos is still a frequently applied management practice, with often negative repercussions for erosion and desertification processes in the area. Indeed, 72% of farmers admitted that they still use fires to clear at least part of their pastures from the thorny bush *Sarcopoterium spinosum* (Beopoulos et al., 2003). However, the same proportion of farmers indicated that they also maintain agricultural terraces that protect olive groves on steep slopes. By contrast, with regard to terraces on pastureland, only 39% of farm respondents said that they performed some activities that helped keep terraces properly maintained.

7.3 CAP-related policy drivers and their influence on landuse changes

Having outlined the crucial role that sheep farming plays in our case study area in north-west Lesvos in the previous section, we now turn to policy drivers that have influenced landuse change in the area and that, at times, have exacerbated desertification problems. Four CAP-related policies stand out with regard to their specific influence (positive or negative with regard to desertification) on landuse change in north-west Lesvos. These include structural policy measures like LFA payments, special policies for aiding small islands in the Aegean, policy measures for the modernisation of holdings, and subsidies paid through the sheep and goat meat CMOs (see also Table 2.1 in Chapter 2).

7.3.1 The importance of Less Favoured Areas policies

As we will show in this section, LFA policies have had a very important impact on landuse in the north-west of Lesvos. LFA policies are the key component of CAP structural measures that have been implemented in the area, and can be split into two distinctive types of policy requirements. The first type of structural measures affects almost all farmers within LFAs, and demands only 'shallow' prescriptions (i.e. not very demanding with regard to changes in farm management) of those farmers wishing to be included in these measures. Most farmers within LFAs are, therefore, able to take advantage of these less demanding structural measures. The second type of measures implemented under CAP structural policies is more demanding of farmers prior to inclusion and only affects specific types of farms.

The first category particularly includes measures implemented under the 'LFA Directive' EEC/75/268 that supports farming in mountain and other disadvantaged areas in the EU. Implemented in 1975, this Directive took effect in EEC member states before Greece's accession in 1981, and member states were asked to provide special subsidies to farmers so that farming activities in disadvantaged areas would be stimulated and living standards of rural communities improved (Villaret, 1996). Due to the geographical, climatic and socio-economic characteristics of our case study area (see Section 7.1), all rural communities in the north-west of Lesvos have been classified as 'less favoured'. The specific aim of LFA subsidies for the north-west of Lesvos has been to ensure the continuation of farming in these particularly disadvantaged areas and, concurrently, to help both maintain a minimum rural population and environmentally sustainable traditional land management systems.

Several amendments have been made to the LFA Directive over time. In 1985, for example, it was incorporated into Regulation EEC/797/85 aiming at improving the efficiency of agricultural structures which, in turn, was superseded by Regulation EEC/2328 in 1991. These latter regulations can no longer be considered special regulations for mountain or disadvantaged areas, although they include various policy measures that continue to specifically target disadvantaged agricultural areas. Among the special subsidies paid through these policies, *compensatory allowances* are particularly important, aiming at providing financial help to farms losing income because of additional production costs linked to both farm location in disadvantaged areas and reduced production due to severe natural disadvantages (Robinson, 2004). In recent years, these compensatory allowances have been particularly implemented through Regulations EC/950 in 1997, EC/1257 in 1999 and EC/1783 in 2003. Farmers eligible for LFA compensatory allowances are those under 65 years of age who agree to remain active in agriculture for five years following receipt of the grant. LFA subsidies are, therefore, aimed at helping disadvantaged farmers in LFAs - such as our farmers in the north-west of Lesvos - achieve equal income levels with non-LFA farmers in order to reduce distortions in competition.

The amount of subsidy is determined in relation to the number of eligible animals and/or the area of eligible crops on the farm, although the list of 'eligible crops' is rather brief in EU LFA policies and does not include intensive cultivation. As we will discuss in more detail below, the fact that livestock numbers define farmer income in LFAs has been an unfortunate policy prescription within the LFA policy arena, as it has often encouraged farmers to exceed the carrying capacity of their pastures. Thus, in the Greek case (since 1991), 'full' LFA subsidies can be obtained by a farmer with livestock numbers equivalent to 20 LUs, and another 50% of subsidy payments can be obtained by farmers stocking additional 10 LUs - thereby encouraging even more overstocking of vulnerable pasture areas. In our case study area, according to statistical data for 1997-1999, each year about 700 farmers were in receipt of compensatory allowances. 94% of the compensatory allowances concerned livestock production, with 95% of eligible land consisting of pastures and only 4% of arable land. LFA payments were for 132 heads of cattle and 78,945 sheep and goats, amounting to a total of nearly 13,000 eligible LUs.

In recent years, there have been two important changes to LFA policies which could affect their environmental impact in the long term. The first changes were linked to the CAP reforms of 1992 and 1999, and particularly to the 'Rural Development Regulation' (RDR) EC/1257/99 where compensatory allowances will be gradually converted to *area payments* over the next few years. The second set of changes is linked to the introduction of cross-compliance for all LFA payments. Thus, under the RDR, farmers must comply with minimal environmental protection criteria in order to receive LFA subsidies. In Greece, this condition implies that farmers have to adhere to the Greek 'code of good agricultural practice'. Our analysis of associated policy documents suggests that current LFA policies increasingly aim at preservation of the natural environment. In addition, the maintenance of farming activity is regarded as equally necessary because of its role in (often sustainable) management of the land, as well as preservation of social life. However, we will discuss in more detail in Section 7.5 how the current subsidy regime, including these LFA payments, has often led to unsustainable pasture management practices in the north-west of Lesvos.

7.3.2 Special aid for small islands in the Aegean

Recognising the special accessibility problems of the remote insular areas of its member states, the Community provides special aid to these areas. In the Greek context, Regulation EC/2019 (implemented in 1993) focuses on islands in the Aegean Sea where special subsidies are provided particularly for small islands with less than 100,000 inhabitants (see also Table 2.1 in Chapter 2). For example, subsidies are given to pay for special supply arrangements for products necessary for human consumption and for the basic means of agricultural production. What is particularly important is the fact that, in our case study area, assistance takes the form of a subsidy for the transport of agricultural products and, crucially, resulting in lower purchase prices for animal fodder. Our survey data shows that farmers in the north-west of Lesvos are heavily dependent on purchased fodder, which accounts for 92% of the total fodder used. In Section 7.5, we will investigate in more detail what repercussions this may have for pasture management and desertification processes in the case study area.

7.3.3 Policy measures for the modernisation of holdings

EU policies for the modernisation of farm holdings are another type of structural measures influencing the development of sheep farming in the case study area, aimed specifically at improving the income of farmers through modernisation of production units. The measures were supported by a specific programme of investments, for the most part financed and controlled by EU and national public authorities. In order to gain acceptance, farm modernisation plans had to fulfil certain economic criteria, thereby making it possible for the future development potential of farms to be assessed. There were two types of measures: subsidisation for farm modernisation and aid to young farmers. Both were based on Directive EEC/72/159 and were implemented in Greece with considerable success from the viewpoint of acceptance by farmers.

According to the initial directive and subsequent regulations on farm modernisation, carrying out projected investments (in animal capital, buildings, machinery) takes place through so-called 'improvement plans'. These policies are not specifically targeted at mountain areas and disadvantaged areas, although they also provide special support arrangements covering farmers in disadvantaged areas. Farms that have already attained a satisfactory level of income, largely by modernising their means of production, are not eligible for agricultural modernisation subsidies, as are farms that, even if subsidised, will not possess the means to reach the targeted income level within a reasonable period of time. On these farms, and in accordance with the underlying 'productivist' rationale of the regulation, non-viable farms should be 'properly' closed down with only social subsidies made available to the head of these farms (e.g. early retirement packages, etc.).

Between 1989 and 1999, a period for which data is available at NUTS[25] 3 level for the whole of Lesvos, 2,408 improvement plans were approved, averaging €35,300 per plan. The number of plans in operation and the amount of average investment were increasing continually over this period, with 43% of the plans approved in 1998 and 1999 alone and with investments amounting to 54% of the total allocated sum in these two years alone. Further, an amount equal to 11% of the total was channelled into investments in animal capital, of which 70% went into sheep and goat farming[26]. In the study area, during 1998, 22 investment plans

---

[25] Nomenclature of Territorial Units for Statistics.
[26] We argue that investment in animal capital is underestimated, as a great part of private irrigation and fencing programmes, amounting to 19% of overall investments, is not classified as 'investment' in animal capital, even though it covers installations for watering and for fencing of pastureland.

were approved, seven of which were submitted by young farmers, and most of which were related to installations for watering of animals and erecting fences around pastures. These data clearly show the vital importance of farm modernisation policies in our case study area - a key driver for further intensification of farming in an environmentally highly vulnerable area.

Implementation of the associated 'aid to young farmers scheme', which aimed mainly at accelerating farm succession, was only implemented in the study area after an eight-year delay compared to the launching of the programme in other parts of Greece. This has meant that the first contract under the programme was only signed in Lesvos in 1994. However, in the third period of implementation of this policy measure under Regulation EC/950 in 1997, there was a mass influx of young farmers into the programme. Thus, of a total of 85 participants, 91% had signed their contract after 1997, and in 1998 alone 18 young farmers received additional subsidy for livestock-oriented improvement plans under this scheme. This sudden upsurge in uptake in can be partly attributed to the desire of local authorities to promote the programme in the livestock-based case study area. This is further emphasised by the fact that only eight beneficiaries (9.5%) received the subsidies for plant production, while all other recipients were subsidised exclusively for livestock-related investments and pasture management.

Three characteristics of these measures are of interest from the viewpoint of desertification processes. The first has to do with the requirement that farms have to be of a specific size in order to be eligible for support (size measured in Annual Working Units [AWUs]). In every agricultural or stockbreeding enterprise, farmers are required to add AWUs subsequent to the implementation of farm development plans. The second characteristic is linked to the lack of encouragement of investments for the purchase of land (such investments are not eligible). Farmers are, thus, encouraged to intensify their production systems, at least insofar as labour is concerned. Finally, even though environmental protection was one of the objectives of EU structural policies (see Chapter 2), and although environmentally-friendly investments were encouraged by the measures in question, it is only recently that there has been any specific reference to the compatibility of these structural policies with environmental objectives through the RDR (EC/1257/99). As a result, an additional 'code of good farming practice' was appended to national implementation legislation, and this code was, initially, identical to that issued in the case of compensatory allowances (see above).

7.3.4 Subsidies linked to sheep and goat meat CMOs

From 1980 onwards, CMOs were implemented for sheep and goat meat (especially Regulations EEC/1837/80, EEC/3013/89, EEC/2069/92, EC/2467/98, and EC/2529/01; see also Table 2.1 in Chapter 2). After 1982, in particular, in order to counterbalance the fall in market prices for sheep meat brought about by the introduction of CMOs, direct subsidies were provided to sheep and goat farmers in the form of an annual premium, with the amount of payments based on the number of sheep and goats owned by an applicant (Boutonnet, 1993). Crucial for landuse change and desertification issues, there were initially no limitations on the number of animals for which subsidies could be received, although in the late 1980s, chiefly for reasons of budgetary constraints (and not for environmental reasons), a ceiling of 500 head was imposed on each farm.

In the EU, it is generally accepted that sheep and goat farming systems are normally situated in physically disadvantaged areas (Robinson, 2004). Thus, a new measure was introduced with a view to counterbalancing the physical disadvantages faced by sheep and goat farmers in LFAs (see Section 7.3.1 above). The above-mentioned ceiling was, therefore, doubled in LFAs so that the maximum number of animals eligible for subsidy now was 1,000 head for each farm. For the same reason, a special system of additional payment per animal was introduced in 1989 for farmers in LFAs. It is worth pointing out that there is a difference between this special aid to sheep and goat farmers in LFAs and the compensatory allowance provided in accordance with the structural regulations highlighted above.

The reform of the CAP in 1992 was based on a programmed reduction in market prices, counterbalanced by direct subsidies, restrictions on the volume of farm production on the basis of individual case history reports on breeding animals, and on a freeze on the area of land under cultivation. The reform involved first and foremost a significant restructuring of CMOs for key crops (cereals, oilseeds and protein seeds). By contrast, however, amendments to CMOs for animal production, particularly sheep and goat meat, comprised only small changes to existing measures, partly because, from as early as 1982, the system for supporting sheep and goat farming based on direct subsidisation of farmers was now firmly entrenched in both bureaucratic systems and farmers' minds. The most important alteration introduced under the 1992 CAP reform for sheep and goat farmers in our case study area was the allocation of subsidy support for each livestock farmer on the basis of reports of individual animal case histories.

As Chapter 2 outlined, in the late 1990s further reform of the CAP market regime introduced basic measures of environmental protection that had to be implemented by member states (Lowe *et al.*, 2002). As a result, requirements for a certain minimum grazing area per LU were established in Greece as a prerequisite for granting subsidies to farms - a move that can be interpreted as foreshadowing the concept of cross-compliance in the sheep and goat meat regime, and a concept enshrined in the 'horizontal' RDR EC/1259/99 (and the more recent Regulation EC/1782/03). In accordance with the requirements of these regulations, in 2000 and 2003 the *Greek Ministry of Agriculture* established, through the above-mentioned 'code of good farming practice', a permissible grazing load which has been set at 'only' 1 LU/ha in island areas. For our case study area this means that, today, only farmers who stock their land at or below this threshold are eligible to receive subsidies through the CMO regime for sheep and goat meat. However, as both Sections 7.4 and 7.5 will analyse in detail, these changes in CMO policies have, only partly reduced grazing pressure on environmentally vulnerable pasture areas in north-west Lesvos.

7.4 Policy impacts on farm survival and livestock farming in north-west Lesvos

Our analysis has so far focused on the development of sheep farming in the study area for the period between 1980 and 2000 (mainly supported by statistical data), and on policies of the same period relating to the development of sheep farming in a region characterised as severely disadvantaged from the viewpoint of natural characteristics. In this section, we will analyse in more detail how above-mentioned policies have influenced farming systems in the area, in particular livestock farming. This will form the basis for understanding impacts of these policies on processes of desertification analysed in detail in Section 7.5.

From the 1970s onwards, public authorities (at first at national level and, after EU accession, at Community level) anticipated that a 'modernisation' process would take place among Greek sheep farmers, leading to increased intensification and specialisation, similar to patterns and processes that had occurred in the beef and poultry production sector a few years earlier. It was believed that intensification of production would be particularly based on a range of techniques, such as genetic improvements or greater use of concentrated fodder in animal diets, which aimed primarily at boosting agricultural productivity. At the same time, it was thought that significant investments in buildings, materials and animal capital demanded by the intensification process would automatically lead to some specialisation of farming and to a more general 'rationalisation' of productive organisation in Greece. Given that economic outcomes of intensification processes are usually closely linked to the possibility of increasing production of animal feed (in order to keep pace with productivity gains), these processes should have also been accompanied by increases in availability of animal feed. Simultaneously, pressures related to type of labour required were also foreseen, linked to increased demand for labour helping with stock handling based on larger numbers of animals. However, this 'modernisation model' encountered significant difficulties in implementation, even in Greek farming areas where conditions were favourable. Rieutort (1995) has emphasised that in mountain and disadvantaged areas, in particular, physical disadvantages of farms created particularly serious, indeed almost insurmountable, problems for this modernisation model. As a result, farm modernisation processes rarely 'trickled down' to extensive livestock-based farming systems in Greece's LFAs.

In our case study area prior to 1970, farming practices that dominated were traditional farming-stockbreeding systems based on plant cultivation, primarily for self-consumption and sheep farming, taking advantage of collective grazing areas by agreement and private pastureland (Giourga, 1991). In this area, limitations on the range of possible farming activities within limits of 'reasonable' competitiveness, increases in productivity elsewhere, and land abandonment processes, all resulted in abandonment of cultivation by mixed farms in particular (Beopoulos, 1997). By contrast, sheep farming was continued relatively successfully and eventually became the prevalent activity. Factors favouring continuation of sheep farming in the case study area were, on the one hand, the presence of relatively low-cost family labour (chiefly parents) and, on the other hand, the presence of pastureland not easily utilised by other types of animals. Moreover, the absence of alternative productive activities obliged many farms to specialise, and, for many, sheep farming was the last resort if they were to remain economically active. Thus, specialisation of farming in north-west Lesvos was often more a result of lack of alternatives than a purposeful choice.

In the sheep farming system of north-west Lesvos, with farming activities that extend over the whole year, increases in labour productivity can rarely take place through intensification of the breeding system. The most feasible route, therefore, has been extension of the farming area, i.e. a shift towards a system based on increases in sheep numbers per farm worker by means of increases in UAA, a squeeze on the cost of intermediary consumption, and simplifications in methods of livestock management. It should, nonetheless, be noted that in disadvantaged areas options for increasing productivity are relatively limited, as farms are already relatively extensive, and reduction of operating loads is often difficult to achieve since they are usually already low. Further, costs linked to the renting of pastures greatly constrict gross margins per head. In this context, it needs to be emphasised that agricultural land in north-west Lesvos is characterised by high degrees of fragmentation, and by large numbers of absentee owners. Indeed, many owners of pastureland live in Mytilini (the capital of Lesvos; see Figure 7.1 above), in Athens, or even abroad, leading to both high percentage numbers of farms renting pastureland and to rented land making up a large proportion of pastures in the case study area. Thus, 60% of farms own less than 25% of the land they use for pasture, and distances between self-owned and rented plots impose new pressures on labour organisation, thereby further significantly limiting productivity.

It is also obvious that there cannot be a limitless increase in sheep numbers. Flock size must be adapted to the characteristics of the productive unit (i.e. the farm), and productivity of labour must be improved accordingly through reduction of farm investments and organisational constraints. Given the severe climatic constraints of the area (see above), farms are effectively kept in existence through year-round utilisation of pastureland and outdoor animal rearing practices. As a result, sheep pens and other buildings for sheep are rudimentary and makeshift to keep maintenance costs low. This is also evident from data provided in improvement plans, as most farms invest in improvement of existing buildings, fencing and watering systems, rather than in new sheep pens. For financial reasons, management of flocks and pastures has also been significantly simplified, with the sheep usually finding food for themselves, and with livestock movement on the various pastures kept to a minimum. As we will discuss in Section 7.5, this latter strategy can be particularly damaging for the vulnerable ecosystems and soils of the area. Thus, in the case study area, given the conditions that prevail, even if incomes from sheep farming are low, this sector is almost the only employment opportunity for farm workers. Most crucially, to get by economically sheep farmers require extra income, and farm subsidies amount to 44% of income at the prefectural (NUTS 3) level and for all farms - a figure clearly illustrating the importance of subsidies for farm survival.

According to our questionnaire survey, factors leading farmers to become involved in agriculture, and in particular sheep farming, are 'family tradition', 'lack of alternative occupations', 'love for animals', and the 'desire to remain in their home district'. According to our interviewees, the latter is only possible if they take up sheep farming. Only once the decision to be livestock farmers has been taken by our respondents would they consider the importance of available subsidies as a second key factor. Thus, although subsidies play a significant role in farmers' decisions to remain in sheep farming, they are often not the original trigger for becoming sheep farmers. Nonetheless, 43% of those questioned argued that financial assistance, either as CMO subsidies or special LFA premiums (see above), is an important factor in deciding to maintain or increase sheep numbers. Another factor which appears to play a role in their decision to stay in farming, is the above-mentioned reduction in production costs secured through subsidies on the provision of animal feed in the context of special measures for small islands in the Aegean Sea. The importance of subsidies is, therefore, obvious in the maintenance of farming activity and, linked to this, preservation of specific land management practices. Without subsidies, it is likely that no agricultural activity would enable the majority of farms in the north-west of Lesvos to achieve an adequate income to secure survival. Subsidies, therefore, contribute greatly towards keeping the rural community in existence. It is, nonetheless, worth noting that not all farming activity in the north-west is subsidy-driven. Our farm survey revealed, for example, that the current favourable combination of factors, such as high milk prices and lower prices for agricultural inputs, has led some farmers to keep animals without subsidisation as an adjunct to subsidised animal capital.

If maintenance of income at some tolerable level can be considered an important outcome of the policy of subsidisation that has evolved since the beginning of the 1980s based on above-mentioned EU policies, it should nevertheless be noted that it would be wrong to see this process as irreversible. The economic situation of farms remains fragile because incomes are still very low compared to EU averages. This means that every restriction on the level of subsidisation, whether from reduction of structural subsidies proposed in Agenda 2000 or from a fall in market prices without compensatory payments, will reinforce the movement towards concentration of production or result in abandonment of farming in the area. Thus, on specialised sheep farms where profits are low, one possible response to withdrawal of subsidies could be further increases in sheep numbers, notwithstanding the technical, structural and economic problems this entails.

As the next section will discuss, even though the establishment of large intensive farm holdings makes continued occupation of land possible, this dynamic would come into conflict with environmental sustainability. Land management in the long term would become precarious and desertification processes - that are already severe in the area (see Section 7.1.1) - would be exacerbated. This, in turn, would mean that there could be no guarantee that the social fabric of the area would survive in its current form.

7.5 Land management practices on sheep farms and implications for desertification

Having outlined the agricultural, environmental and policy-related situation in the case study area of north-west Lesvos, in this section we wish to analyse in more detail the *interlinkages* between policy drivers for landuse change and their implications for desertification processes. Based on our field research in the area, information provided by the farmers themselves and information based on Ashworth (2000), farming practices thought to have the most important impacts on desertification processes include, in particular, flock management practices (including size of flocks and flock movement), maintenance of terraces, and use of fire. Table 7.2 shows the predominant farming practices used in our case study area, and highlights that all of these key farming practices are important in our case study area (in a positive and negative sense). Thus, use of fire is still very common in the area (63% of respondents) and nearly the same amount of farmers (62.5%) regularly move their sheep flocks - both with positive and negative repercussions for desertification (see below). Maintenance of terraces, meanwhile, shows a mixed picture, with terraces in olive groves being regularly maintained by over two-thirds of respondents, while terraces on pastureland are maintained by less than 40% of farmers - the latter a crucial shortcoming on land that is already prone to desertification processes.

In the case study area, as in many other Mediterranean livestock farming areas (see also Chapters 4-6 in this book), the frequent use of fire as a form of 'traditional' land management practice, has particularly negative repercussions for desertification. Farmers use fire on pasture areas to encourage the growth of new tender shoots for sheep grazing, but this has engendered an environmentally unsustainable vicious circle. Regular fires have resulted in the gradual predominance of an undesired thorny bush (*Sarcopoterium spinosum*) on pastures which is unpalatable for animals. The necessity for destroying this bush, which quickly negatively affects pasture quality, forces farmers to light fires even more frequently, exacerbated by the high cost of alternative means of getting rid of this bush, such as chemical methods or mechanical uprooting. According to Papanastasis (1977), this 'self-imposed' need for the continuation of regular fire use has led to further destruction of the last remnants of protective vegetation cover on many pastures in the area, with resulting aggravation of erosion.

Beyond the use of fire, the pressure exercised by grazing animals on the capacity for renewal of plants on pastureland is also vital to keep erosion processes under control. This grazing pressure is customarily expressed through the 'grazing load' indicator, i.e. the number of animals per unit of grazing land. Ashworth (2000) suggests that grazing load can be regarded as a suitable indicator for showing environmental pressure. Nevertheless, the reliability of this indicator has been criticised, especially as it does not take into account seasonality in grazing. In the insular regions of Greece, and in our case study area, pasture areas are of limited grazing capacity and, therefore, insufficient for covering the animals' nutritive requirements for the whole grazing period. The natural outcome of this imbalance is overgrazing (Papanastasis and Giannakopoulos, 1989; Papanastasis, 1998). Yet, no studies

Table 7.2: Farming practices used by respondents in north-west Lesvos (Source: Beopoulos et al., 2003).

Farming practice	Yes	%	No	%
Fire use	63	63.0	27	27.0
Movement of sheep flocks	60	62.5	36	37.5
Maintenance of terraces in olive groves	31	72.1	12	27.9
Maintenance of terraces in pasture areas	28	38.9	44	61.1

have been conducted on overgrazing in the case study area, and no specific threshold has been established above which grazing load may accelerate desertification. We calculated the grazing load (based on data provided through our farm survey) as 0.68 LU/ha, taking into account only pasture areas. But grazing also occurs in areas under cultivation, especially under olive trees, and if these areas are also taken into account, the grazing load falls to 'only' 0.56 LU/ha - well short of the maximum grazing load of 1 LU/ha suggested by the code of good farming practice (see above).

Table 7.2 (above) highlights that over 60% of farmers regularly move sheep flocks between pastures. The significance of such flock rotation is that, if practised in suitable ways, it enables the renewal of pasture plants. Yet, our farm survey indicated that flock rotations were only taking place for purposes of flock management (on the basis of age, sex and animal productivity), and not for reasons of sustainable management of pastureland. Lack of flock movement results in constant presence of grazing animals in the same area, frequently in large numbers, with consequent overgrazing and resultant trampling and destruction of new shoots that could help prevent erosion. As indicated above, in disadvantaged areas flock rotation practices are determined first and foremost by increases in sheep numbers per worker, with simultaneous expansion of pastureland. In the study area, land ownership patterns linked to absentee landowners (see above), and resultant lack of available pastureland for renting, have resulted in increases in the grazing load, as farmers have no access to areas where they could put their additional sheep. The willingness to embark on costly sheep rotation practices is also affected by a squeeze on the cost of inputs. In their endeavour to deal with this pressure, sheep farmers resort to purchasing subsidised concentrated animal feeds that are cheap but of limited nutritional quality, resulting in further grazing pressure on the scant areas of pasture available. Further, in order to reduce pressures imposed by the large amounts of strenuous work involved in the everyday running of the holdings, many farmers also simplify their land management practices by cutting down on flock movements.

Pressures on environmentally-vulnerable pastures are further exacerbated by the neglect of traditional environmentally more sustainable farming practices that were common before 1970. In particular, in a region such as north-west Lesvos characterised by steep gradients and eroding hillsides, terraces have played a crucial role in protecting cultivated land from erosion. As in the other case study areas analysed in this book (see Chapters 4-6), these stone-built terraces are a characteristic element of the island landscape. However, the 'forced' specialisation in sheep farming over recent decades (see above), and the associated abandonment of cultivation, have resulted in a diminution in the importance of terraces in the present-day production system. No longer comprising a basic functional element in sheep farming, terraces have been neglected and are beginning to collapse. The economic costs involved in repairing them, given present-day realities, are an obstacle to their restoration. As a result, and as Table 7.2 (above) highlights, 61% of our survey farmers said that they no longer maintained terraces on their pastureland, with chief reasons being both costs (monetary and time) associated with terrace maintenance and the fact that they could see no reason why they should maintain these terraces. The latter is a particularly worrying answer for the prospect of desertification alleviation in the area, as it suggests that farmers have currently little awareness of the role played by their terraces in preventing erosion.

However, the fact that a significant proportion of UAA in the area is rented, usually on yearly or shorter-term contracts, often prevents farmers from engaging in long-term investments and land improvement initiatives, such as maintenance of existing terraces and construction of new ones.

7.6 The local dimension of policy promotion

We have, so far, highlighted the importance of both land management practices and the policy environment in shaping landuse practices in the case study area. However, as many studies have highlighted (e.g. Buller *et al.*, 2000; Wilson and Hart, 2000), the existence of policies does not necessarily mean that these policies will be implemented successfully on the ground. In this last section of our analysis, we, therefore, wish to examine how policies and measures favouring the development of sheep farming were implemented in the study area. We will place specific focus on the mechanisms that have underlain state intervention for many years, and through which policies have been promoted to the farmers. We then go on to outline how these measures have been perceived, utilised and evaluated by the farmers themselves. In order to reconstruct the way in which measures were applied, we administered questionnaires to, and held interviews with, livestock farmers, and also carried out in-depth interviews with officials in regional, prefectural and local government, as well as with a representative of the development agency of Lesvos (see also Section 7.1.2 and Chapter 3).

7.6.1 The role of institutions in charge of policy implementation

As Chapters 8 and 9 will discuss in more detail, policy-making in Mediterranean countries is often characterised by instrumentalist approaches based on top-down decision-making structures that, in many cases, give little room for non-state stakeholders to be both heard or actively involved in the policy implementation process. Having outlined the precarious situation of farming in our case study area in Lesvos in the previous sections, this section will briefly analyse the wider motivations and rationale behind a Greek policy philosophy that continues to emphasise productivist development over environmental conservation, despite increasing evidence that some agricultural systems are increasingly farmed in environmentally unsustainable ways, leading to further desertification.

Top-down procedures of policy formulation and implementation are typical for all policies under investigation in this chapter. This situation is often exacerbated by the fact that almost all policies that have been identified as influencing landuse change in our case study area, and thereby affecting desertification, have been developed at EU level and are, therefore, relatively 'removed' from local decision-making actors. Although some limited national input was allowed with regard to influencing levels of subsidies to be paid to farmers for certain EU policies (e.g. CMO for sheep and goat meat, LFA subsidies, and financial aid provided for small islands of the Aegean), interventions at the Greek national level have been limited, relating largely to favoured treatment for special population groups (young people) and regions (LFAs and islands). Nonetheless, more intensive intervention by the *Ministry of Agriculture* occurred with regard to the two structural measures requiring some input on the part of the producer, i.e. improvement plans and plans required to enable young farmers to take over farms (see above). In this context, and following the logic of the structural plans, the *Ministry of Agriculture* made a series of choices at central level on matters pertaining to minimum and maximum size of farms eligible for subsidy, proportional funding for each individual investment (e.g. for water saving or investments in equipment purchase), and establishing levels of subsidies between different regions or farm production types. It should be highlighted, however, that the general orientation of policy implementation occurred in accordance with the continuing predominating productivist views at the *Ministry of Agriculture*, which has led to an emphasis on investments in farm modernisation and, above all, improvements in farm productivity, and not on investments in environmental protection.

The differential treatment between the two categories of support (i.e. market versus structural) that was pursued at the policy formulation stage was continued at the stage of on-site implementation. For the three 'general implementation' measures highlighted above, the role of local agronomists and the *Directorate of Agriculture* in Lesvos was restricted to simple bureaucratic processing related to receipt of applications and necessary documentation, their dispatch to the service headquarters, announcement of names of beneficiaries, and the amount of support received. A similar role was played by the cooperative in Lesvos in the purchase and distribution of subsidised animal feeds and assistance to farmers in dealing with bureaucratic formalities. Our investigation has revealed that at no point in time where these institutions actively engaged in shaping the policies they were administering or, more crucially, in questioning the possible impact that these policies may have in exacerbating desertification processes in already highly eroded livestock farming areas.

In the case of on-site implementation of improvement plans and support for young farmers, the Lesvos *Directorate of Agriculture*, together with agronomists in the local branch office, had a larger role to play, especially as the requirement for a decision to be taken on enrolment or non-enrolment of farmers in the support programmes necessitated involvement on their part. It is also worth bearing in mind that private agronomists in most regions of the country were actively involved in the drafting of improvement plans (in some cases their participation was compulsory), and so agronomists made a significant contribution to briefing producers on the programmes. It is, nevertheless, clear from our findings that their contribution to information-provision for policy-recipients (i.e. the farmers) was limited.

With regard to monitoring and evaluation of policies implemented in north-west Lesvos, office-holders in local development organisations, as well as associated 'intermediate' stakeholders, only had a limited role to play in policy implementation processes. In practice, their involvement was confined to dispatching data (in ready-made format moreover!) on uptake and use of subsidies provided through the policies, while the ex-post evaluation of policy impacts in Lesvos, carried out in the 1990s, was the work of outside assessors relatively 'removed' from the situation and problems on the ground. To a large extent, this 'policy implementation vacuum' at local level explains why Lesvos farmers had relative freedom to use policy subsidies to the maximum, without being encouraged to question the rationale behind these policies and their possible impacts on environmental resources (i.e. the last remnants of protective plant cover) on their farms.

7.6.2 Farmers' perceptions of policy implementation mechanisms

How did the farmers themselves perceive the mechanisms for implementation of policy measures, and did they at any time question the policy pathways chosen by external actors to support their farming activities? We have seen in above discussion that the EU policy-based subsidy system in the north-west of Lesvos has been attractive to all farmers, even if the amount of subsidy received is often considered inadequate for ensuring farm survival. It is for this reason that farmers are keen to maintain good relations with the institutions responsible for implementation of the measures and, accordingly, for payment of subsidies. The presence of a variety of institutions in charge of policy implementation (see above) increases the potential for farmer contacts, as well as contributing to the development of various privileged relations between some farmers and institutional actors. It is important to highlight in this context that the development of farmer relations with institutional actors is not only based on economic benefits derived from subsidies, but also on opportunities offered by obtaining information and by giving and receiving advice.

The above-mentioned specific policy implementation structures in existence in Greece have enabled certain actors in the implementation process to have relatively large amounts of power. Here, the role of agronomists (at least those who are part of the local extension service) is particularly worth mentioning, as they have become indispensable consultants for sheep farmers in north-west Lesvos, imparting both information and advice. The technical experts, mostly agronomists and veterinarians well acquainted with the area, appear willing to listen and to discuss problems with farmers, transmitting their demands to other authorities, and so becoming privileged interlocutors in the policy implementation process. Table 7.3 shows relations between farmers and agencies involved in policy implementation in north-west Lesvos, based on multiple answers provided by farmers to questions about who

Table 7.3: Relations between farmers and stakeholder groups involved in policy implementation in north-west Lesvos (Source: Beopoulos et al., 2003).

Institutions/persons	Consulted	Ranking order	Comprehend problems	Ranking order	Inform	Ranking order	Solve problems	Ranking order
Colleagues	55	(1)	25	(2)	24	(3)	4	(4)
Local extension officers	24	(2)	27	(1)	15	(4)	34	(1)
Family	15	(3)	4	(4)	3	(6)	4	(4)
Directorate of Agriculture	9	(4)	14	(3)	31	(2)	24	(2)
Cooperative	4	(5)	3	(5)	44	(1)	14	(3)
Municipality	1	(6)			10	(5)	2	(6)
Farmers' association	1							
Prefectural government							3	(5)
State (Ministries-Services)								
Scientists								
None	10		27				19	

provides them with advice, who understands them and their problems, and who provides them with information to solve problems. The table shows that agronomist working for the local branch of the extension service appear to be closer to farmers than anyone else involved in the implementation of policies affecting sheep farmers. Local agronomists are turned to chiefly for advice, for the understanding s/he shows towards farmers' problems, and for problem solving. They are, nonetheless, used less for obtaining new information, where the local cooperative appears to be the main point of contact.

The *Prefectural Directorate of Agriculture* plays a central role in promoting and implementing EU agricultural policies, while the local extension service provides back-up for concrete implementation of actions and programmes. For this reason, the latter are perceived by farmers as a place where reliable information can be obtained rapidly, and where problem solutions can be found. By contrast, the advisory function of officials working in the *Prefectural Directorate of Agriculture* is of limited significance because, given existing levels of organisation, they are neither able to establish relations of trust, nor a feeling of 'complicity' with farmers comparable to that offered by agronomists working in local branch offices. Another reason for the possible breakdown in trust between the *Prefectural Directorate* and the farmers of north-west Lesvos is that the Directorate has its headquarters in the island's capital (Mytilini), about two hours' drive from the study area. Our results show, therefore, that trust in policy advice is largely a local matter and that knowledge of farmers' problems is key for farmers' acceptance of 'official' policy advice.

This is a crucial point with regard to understanding the relatively uncritical acceptance of farmers of subsidies that, eventually, will lead to further environmental degradation in their area (see Section 7.5). Indeed, the importance of subsidies as a vital part of farm income makes information and its effective dissemination a factor of the utmost significance for farmers. As Table 7.3 (above) shows, the local agricultural cooperative is, according to answers from the farmers in our sample, the most important information source in this respect. The granting of subsidies for animal feed, the provision of assistance in dealing with bureaucratic red tape through the cooperatives and through the - *de facto* - frequent contacts with the cooperative management, and the relationships of confidence and intimacy that can grow between farmers and staff from the cooperative, may all go some way towards explaining why farmers tend to turn to the cooperative for information (see also Hagedorn, 2002). The fact that the cooperative is also often a long established institution, together with farmers' perceptions of effective management by the cooperative of farming business (especially marketing of milk and manufacture of cheese), also helps to explain the high esteem that the cooperative has in the eyes of the farmers with regard to problem-solving issues. This means that farmers in the north-west of Lesvos have, so far at least, relatively uncritically accepted subsidy advice given to them by the cooperatives - irrespective of whether this leads to further desertification on their farms or not.

It should be noted that the farmers in our sample, contrary to initial expectations, did not make any reference at all, or at least referred only obliquely, to the role of the municipal government, the regional administration (for the Northern Aegean), or local action groups as part of the EU LEADER I initiative as possible key stakeholder groups in the provision of advice and information. Three reasons emerge from the farmer interviews. First, farmers argued that the role of the municipal government (at NUTS V level) is limited, and that it is seen as making only a small contribution to keeping farmers in the north-west of Lesvos informed. This finding is difficult to reconcile with the fact that Community authorities have appointed a special team of agronomists to deal with problems of municipal residents, for the most part livestock farmers themselves. A possible explanation for the lack of involvement of officials from the municipal government is that appointments have been recent, and that staff have not had time to 'bed down' and show that their institution could provide information to farmers effectively. Second, contact to regional (NUTS II) authorities is also weak. Although formally an important actor in policy planning and implementation, particularly of structural policy (aid to young farmers and improvement plans), it seems that the *Regional Administration for the Northern Aegean* has not established contact with farmers in north-west Lesvos. Again, a possible explanation for this may be the fact that, compared to the *Prefectural Authority*, the regional administration is a relatively new administrative structure. Third, it is also interesting to note the absence of farmers' reference to the *Local Action Group*, in operation on Lesvos since the beginning of the 1990s largely as a result of the LEADER I initiative. Despite the fact that this group has organised seminars on rural tourism, that it has promoted investments in both rural tourism and the processing of agricultural products in the study area, and that it represents itself as the institution that 'grounds' its activities in consultation with the local community, it does not appear to have made much headway with the farmers.

Overall, our investigation of factors determining the frequency of contacts between farmers and institutional mechanisms for policy implementation in relation to sheep farming has yielded, more or less, the expected results. Large farms (from the viewpoint of economic size) usually develop multiple relationships with the agencies involved in policy implementation. In particular, young farmers with well differentiated farms in terms of techno-economic orientation tend to discuss their problems with appropriate officials and turn to them for help. Similarly, sheep farmers who are very active on their holdings, as shown by the size of their flocks, level of investment, area of rented pasture, volume of animal feed purchased, and amount of family labour utilised, are those most on the lookout for information.

Analysis in this section has, therefore, shown that farmers turn mostly to three administrative institutions as a way of solving their problems and acquiring the information they need with regard to subsidies and policy implementation: local agronomists, the *Prefectural Directorate of Agriculture*, and the local agricultural cooperative. However, these are institutions where, as discussed above, the flow of information is in one direction (i.e. top-down), with minimal participation of stakeholders in planning and application of policy measures. Decision-making, as interviewed stakeholders themselves acknowledged, takes place elsewhere, and all actors (including those that farmers identify as participants in policy-making) see themselves as mere implementers of policies designed elsewhere.

7.7 Conclusions

This chapter has highlighted a variety of key points with regard to understanding the interlinkages between the policy environment and desertification processes in our case study area of north-west Lesvos. We have shown that in the study area between 1981 (year of Greek entry into the EEC) and 2001, the population figure has remained relatively stable, with a slight decreasing tendency, and that sheep farming remains the only farming activity and the principal economic activity in the north-west of Lesvos. However, the case study area is also slowly losing its importance for sheep farming on the island, as testified by developments in animal numbers, which - for the same period - have increased faster in other parts of Lesvos. The chapter has particularly shown that subsidies received by farmers, either through the support system for sheep farmers (CMOs for sheep and goat meat) or through EU structural measures, has contributed to the continuation of farming activity and the retention of population in the case study area - a situation different from many other parts of the country where rural areas have continued to decline. On the positive side, the policy record has, thus, been one of achievement of EU policy objectives to retain a certain minimum population in some disadvantaged areas and, aided by this process, the maintenance of agricultural practices and avoidance of the rural exodus and abandonment of land characteristic for so many other regions of the Mediterranean (see Chapters 4-6). It is obvious that other factors have also contributed to population retention, such as the demand for sheep milk on the local market and the manufacture of cheese (as mentioned above, three protected 'denominations of origin' cheeses are produced on the island), together with limited employment opportunities elsewhere on the island and beyond.

These factors have contributed to the establishment of a farm production system conditioned by the natural disadvantages of the area. An extensive sheep farming system was the logical farm development pathway for many farmers in north-west Lesvos, similar to farming patterns prior to 1970. This fitted in well with the promotion of extensive pastoral systems as one of the goals of EU policy. However, in order to increase farm profits, north-west Lesvos sheep farmers adopted a farm development strategy based on a system predicated on raising the number of sheep per worker through increasing the amount of UAA - a system aided and encouraged by various EU policies over the years. This, in turn, squeezed input costs and simplified flock management. Yet, in an area of high sensitivity to desertification, this practice led to increased danger of desertification. There were two key reasons behind this issue that could not be foreseen by those devising policies that affected north-west Lesvos farming landscapes. First, the prevailing structure of land ownership in the study area (mainly absentee ownership leading to large proportions of rented land) precluded holdings from acquiring significant areas of pasture. The resultant grazing load, although within the limits prescribed by the Greek 'code of good farming practice', has substantially exceeded the carrying capacity of the meagre pastures and has led to visible and measurable exacerbation of desertification processes (see Geeson *et al.*, 2002). Second, because certain traditional practices of pasture management have been abandoned, as they are no longer consonant with the logic of the productive (and productivist) system, the dangers of erosion have further increased. As a result, movement of sheep to different pastures has been minimised and, even more crucially, terraces are no longer maintained and collapse. These negative land management practices are further exacerbated by regular fires used as a tool to eradicate unwanted vegetation.

The prospects with regard to future desertification alleviation through changed land management practices and a changing policy regime, therefore, look relatively bleak. Even though environmental protection was one of the aims of the EU structural regulations, and although environmentally-friendly investments were encouraged by many of the EU policies discussed in this chapter, it is only recently that specific reference has been made to the environmental compatibility of implemented policies through the new EU RDR (EC/1257/99). Yet, only time will tell whether the suggested changes to the CAP, as envisaged under Agenda 2000 and put into action with the 1999/00 and 2003/04 reforms, will bring long-lasting environmental protection to the north-west of Lesvos. The situation is currently certainly not aided by the fact that the prevailing general view in Greek public administration continues to be one that does not include much concern for the environment. Indeed, the need for keeping rural populations, and the modernisation of farm units, continue to be the general context within which policy implementation is carried out in the productivist Greek agricultural policy framework - with potentially damaging effects for fragile pasture areas in the north-west of Lesvos.

Part III

Desertification, policies and actor networks in Southern Europe

Chapters 4-7 have given us ground-based empirical evidence of the often negative interlinkages between the national and EU policy environment and desertification processes in our four Mediterranean case study areas. Building on this discussion, the final part of the book will analyse in detail the complex interconnections between desertification, policies, and actor networks in Southern Europe. In this part of the book we will adopt a more theoretical approach that is aimed at helping us understand the subtleties behind policy and desertification interconnections. To this end, **Chapter 8** will first provide a detailed justification for the use of *actor-network theory* through a contextualisation of the actor-network approach as part of a post-modern interpretation of societal change. **Chapter 9** will then investigate in detail actor networks and the implementation of policies affecting desertification in Southern Europe by paying particular attention to actor networks and power in policy implementation, by identifying discourses of desertification and policy agendas at the local level in our four case study areas, and by analysing how these agendas influence the interfaces between policies and desertification in Southern Europe. The conclusions to the book (**Chapter 10**) will bring the different strands of our investigation together by presenting a synopsis of how our study may contribute towards global and international debates on desertification, and by providing recommendations for future desertification mitigation in Southern Europe.

8. Actor networks, policies and desertification: some theoretical considerations

M. Juntti and G. A. Wilson

We argue in this book that the impact of policies on desertification cannot be fully understood when treated separately from the complex socio-economic context of policy implementation. As discussed in Chapters 2 and 3, the goals of policies, such as the CAP CMOs, the Agri-environmental Regulation and the RDR, evolve throughout the process of policy implementation. Although EU-related policies formulated at the supra-national level are not explicitly open to interpretation and modification in the implementation process, they are nonetheless both 'translated' into practice through a process of deliberation and negotiation at different scales and mediated by institutional and cultural contexts. As a result, and as we have seen in our case study discussions (Chapters 4-7), such policies can be adapted to serve various agendas that may become prevalent in land management in a specific locality. Further, and as Chapter 2 highlighted, by focusing on policies relevant to desertification, the UNCCD and the subsequent NAPs prescribe an inter-sectoral and participatory approach that has implications for action and interaction at various scales of administration and land management. This means that the plethora of involved interests, interpretations and discourses, as well as the structures and processes of policy implementation, are vast and varied and need to be better understood if we want to get as full a grasp as possible of how policies may act as driving forces behind desertification processes in Southern Europe.

This forms a challenge for both practical and theoretical aspects of how best to integrate desertification mitigation goals into land management decision-making - a key issue we wish to address in this chapter. While Chapter 9 will address these issues on the basis of data from the case studies in Southern Europe, this chapter will discuss the broader theoretical and conceptual premises and implications of an interpretative, actor-oriented approach to governmental intervention in environmental governance more generally, and rural development in particular - two key issues affecting desertification in our four case study areas. To this end, Section 8.1 will investigate interlinkages between actor-oriented approaches and the newly emerging concept of 'governance'. Section 8.2 will then delve more deeply into theoretical considerations concerning actor-network theory (ANT) as part of post-modern interpretations of societal change. Section 8.3 will conclude this chapter by arguing that only through a combination of ANT and 'new institutionalist approaches' will we be able to better understand the role of actors in policy-making and implementation affecting desertification in our four case study areas.

8.1 In search of new concepts of environmental governance

As discussed in Chapter 1, evolutions in technology and associated risks, globalisation of knowledge and the economy, and the subsequent demise of many modern institutions of power, have developed more or less in parallel with 'post-modern' theorisations of society. The actor-oriented approach predicated on ANT used in this book is associated with a post-modern view of the world, and can be contextualised under a general heading of 'post-structuralist' approaches which emphasise the role of human perception as an organising force, while not accepting the presumption that interpretation rests on a pre-determined, rigid structure of meanings (see also Section 1.4 in Chapter 1). The abandonment of structural explanations reveals the politicisation of reality, which is represented, maintained and acted upon in a 'discursive praxis', where discourse both maintains and conditions a certain understanding of the world. As Prior (1997) emphasises, this establishes and maintains the phenomena or practices, which empower (or disempower) some agents to impose their interpretative frameworks over others. According to Foucault (1991), discourse restricts, limits and arranges what can or cannot be said about specific phenomena and, thus, has an impact on how knowledge comes to be produced, encoded and displayed. We argue here that these are forces that shape how decisions about natural resource management are made, and how these resources are represented in the policy process. In practice, social and political contexts offer interpretative methods for social realities, with some orientations being privileged over others, although this often depends on the institutional setting of the interaction. All strategies and constructions used by actors are, therefore, drawn from available discourses and are, to some degree, simultaneously shared with others or adopted from antecedent practices.

These debates are closely interlinked with recent theorisations in the field of *rural sociology* that have had important repercussions of how to conceptualise land management decision-making by various stakeholder groups in rural society. In rural sociology, these theorisations are particularly visible in the paradigmatic shifts of the last decades. In the 1960s and 1970s, for example, conventional approaches began to be subjected to critiques concerning their capability to address political questions like rural poverty and deprivation. In particular, Pahl's (1984) demolition of the traditional conceptual framework of rural studies based on the 'rural-urban continuum' in the 1960s called for a new definition of the 'rural' altogether (see also Hoggart, 1990). While mainstream focus shifted more to agrarian questions, addressing rural class divisions and the question of family farm persistence in the 1970s, the 1980s and 1990s brought processes like trade liberalisation, globalisation and the transition towards post-productivist or multifunctional agricultural regimes to the forefront, which further undermined old perceptions about the nature of modern society. This was perceived as a process of restructuring by contemporary rural social scientists, whereby the 'restructuring approach' served as an attempt to fill the 'theoretical vacuum' initially created by the earlier critique (Newby, 1980; Buttel, 2001). As a result, rural issues became more complicated with the increasing mobility of capital, the rise of environmental and consumer interests and the de- and re-regulation of political and economic structures. This led to an emphasis on bottom-up approaches in rural sociology, granting land management stakeholders, such as farmers, the role of a knowledgeable political actor, and subjecting farm-level decision-making to closer scrutiny. In this context, the Dutch 'Wageningen school of thought' can be seen as home for leading theoretical thinkers of actor-orientation in rural sociology (e.g. Van der Ploeg, 1997; Long, 1997).

Key for our investigation of policy and desertification interlinkages is that once farmers (or other rural actors) are treated as the *mediators* of external influences that pass through the farm gate, the distinction between the public and private domains in politics becomes increasingly blurred. This process has also inspired a shift in academic enquiry from a focus on *government* to various mechanisms of *governance* (Frouws and Mol 1997; Wilson, 2004). In rural geography, in particular, these shifts have led to a cultural/reflexive turn, depicting a growing sensitivity to the immaterial, as well as the political, implying not only an expansion of the boundary of geographical imaginations and a curiosity towards the 'rural other', but also a different conceptualisation of power (Phillips, M., 1998). Yet, as the discourse and cognitive structures became recognised as rules and resources (i.e. organising features) of society, their interrelations and localised histories began to require more analytical attention, as did their relation to 'the material' which still remains far from salient. Thus, regardless of disciplinary affiliation, the consequent challenge to social science inquiry to rural issues can be described as one of conceptualisation and explanation. According to Marsden *et al.* (1993, 1), "the contemporary processes of global restructuring emphasise how the political, the social and the economic interact within and between different local, national and international spaces. No one process is dominant. These tendencies challenge all aspects of social enquiry."

One specific question raised by the notion of restructuring has concerned bridging the gap between theory and empirical enquiry. The processes of economic and political restructuring, and the integration of environmental concerns into policy processes and decision-making, take effect at various different scales and across various sectors, both in the private and public domains, and need to be studied through thickly textured, context-sensitive empirical enquiry, and organised by concepts that span both structure and agency (Adger *et al.*, 2003). In this context, the *actor-network approach* is one concept that provides a response to Pahl's initial contesting of the traditional economic forms that political economy approaches saw as the organising factors defining the concept of 'rural'. Directing focus to the interaction and links between actors not only at material but also at interpretative level, the actor-network approach expands the actor-oriented focus from farm-level considerations to a much wider set of stakeholders and their interrelations (Buttel, 2001). This *constructionist* network approach now holds increasing currency in social science research into rural issues, addressing through conceptual analyses local action and non-material considerations and their role in ongoing transitions. Thus, relying on the idea of interpretative power and the crucial force of human perception in the organisation of society, the actor-network approach draws attention to the relationships or links between actors which, in turn, helps explore and understand action in localities. Examining the concepts of 'locality' and 'rurality', Marsden *et al.* (1993) claim that it is these actor links in particular, rather than geographical factors, that define localities.

We argue in this book that this approach should constitute a significant element in the inquiry into the governance of natural resources and the policy dimensions of desertification. The idea of a discursive praxis, consisting of "the existing institutions and power relations governing and governed by social action in the form of either verbal expression or social practice" (Long, 1992, 26) and applied to the research of governance, means that attention should be directed at factors of administrative culture, the processes and struggles of interpretation, and involved actors as well as formal institutions, in order to understand policy outcomes and impacts on phenomena like desertification.

As discussed above, concern and demand for environmental sustainability form a significant force within changes in rural and agricultural restructuration, particularly at a European scale. Although decentralisation, local empowerment and participation are often associated with environmental sustainability, and can be seen as crucial factors in increasing policy ownership and adoption, it is by no means self-evident that environmental sustainability - i.e. in our context through desertification alleviation or mitigation - is guaranteed by the recent transformations or indeed by any of the above processes alone. Ironically, and as Chapter 9 will analyse in detail, the changes in (or into) rural governance seem to further complicate the integration of environmental concerns into natural resource management in rural areas. Not only is the plethora of actors and agendas steadily increasing, but - as depicted by the example of desertification mitigation by means of the NAPs discussed in Chapters 2-7 of this book - transposing objectives formulated at national, European or even global level to local circumstances provides a challenge in itself. Therefore, the issue of understanding the mechanisms, including barriers and opportunities, of environmental integration into rural governance in Europe remains pressing. As both Jones and Clark (2001) and Adger *et al.* (2003) highlight, the multi-level and cross-sectoral domain of rural development, bringing together a plethora of policies, actors, agendas and strategies, forms an increasingly challenging arena for issues like desertification mitigation.

While the theory of a 'Mediterranean Syndrome' (see Chapter 3) emphasises structural features of environmental governance in Southern Europe, in Chapter 9 we will use elements of discourse analysis and the actor-network approach to conceptualise the interpretative dimension of 'desertification governance' in our case study areas. As discussed in Chapter 3, our analysis is also guided by 'new institutionalist' theorisations drawing attention to the processes, strategies and mechanisms of power, interests and knowledge in the European multi-level governance system. As a result, our analysis of the links between desertification and policies in Southern Europe focuses on the actor networks of land management and desertification in our case study areas, as well as on strategies and practices of communication that influence the transmitting and exchanging of information concerning goals and policies governing land management. While our analysis of the case study material does not exploit the full analytical capacity of the ANT, our main aim is to establish a basic understanding of the interpretative dimension of environmental governance in Southern Europe, with specific focus on desertification. Simultaneously, we also wish to begin to develop a set of concepts that will integrate post-structuralist approaches to policy analysis and evaluation more widely.

8.2 Contextualising actor-network theory as a part of post-modern interpretations of societal change

In this section we wish to discuss the specific characteristics of ANT that render it a particularly useful approach for addressing environmental governance, while also outlining some weaknesses that call for the support of additional explanatory theories of governance. This section is structured into three parts. First, we will look more closely at ANT as an emerging concept for better understanding human-nature interactions (Section 8.2.1). In Section 8.2.2 we then investigate the interlinkages between ANT and the politics of nature, while Section 8.2.3 looks specifically at how ANT may help us to theorise the power of specific actors in complex multi-layered stakeholder hierarchies.

8.2.1 Actor-Network Theory: towards a new understanding of human-nature interactions?

The aspect of conventional political economy approaches to rural issues that attracts most criticism is both the tendency to dichotomise (especially related to the dichotomy of society and nature) and the fact that it emphasises structural drivers of human action (Marsden, 2000, 2004). However, the cultural turn and post-structuralist analytical approaches have led to a reconsideration of the concept of 'nature' (Castree and Braun, 1998; Goodman, 2001). The exact status of the 'physical reality out there' remains under debate, but the main implication of the post-structuralist constructivist approach to nature is the *politicisation* of how and what meanings we attach to the concept of nature. This obscuration of the spheres of 'the natural' and 'the social' not only dissolves the dichotomy, but also corrodes the status of scientific knowledge as the guiding light of environmental management and forces us to pay attention to the normative and/or moral statements and beliefs inherent in how nature is represented in these debates.

ANT is one approach resulting from this reconsideration of science and nature. It is particularly useful in overcoming the rift between society and nature, and remains sensitive to the moral implications of environmental management discourses. In line with the post-structuralist tradition, ANT maintains that the concept and role of nature cannot be defined 'a priori' without having outlined the network of relations in which constitutive exchange of properties can take place. As Section 8.2.3 will highlight, this also has significant implications for the conceptualisation of power. As Goodman (2001) argues, the concept of agency (as well as that of structure, society and nature) is an outcome of network building and maintenance and can be described as socio-material in character.

The ANT, deriving from the anthropology of 'technoscience' of Bruno Latour, has been classed as one of the leading four foci of theory and research in the late twentieth century sociology and political economy of agriculture (Buttel, 2001). Aiming to bind culturalist dimensions into the study of power, and providing a detour around the impasse of macro- versus micro-analysis, ANT presents itself as a credible approach to complement the conventional political economy approaches in rural research (Phillips, 2002). Other rural researchers have emphasised the current tendency to dichotomise the present social science field of rural studies, dividing it into two opposite camps, that of structuralist political economy and the other of post-structuralism (Hoggart, 1998; Abram, 1998, 1999). Yet, ANT as both a theoretical and methodological tool is not altogether unproblematic. While the relativism inherent in ANT allows the consideration of a heterogeneous array of actors - human, inhuman and immaterial - and, thus, provides a very useful and novel 'view' into the perceptions and relationships governing the issue under focus, it renders the theory less powerful as a model of explanation of cause and effect (Callon, 1999). Therefore, ANT, although clearly emerging from a post-structuralist background theory of power and discourse, challenges this paradigmatic dualism and "has been ... seen as a means of adding more contingency, indeterminism and process to political economy perspectives" (Phillips, 2002, 83).

However, as we argued in Chapter 1, the increasingly significant role of human perception as an organising factor of society, linked to the demise of structural manifestations of power and of the unquestioned authority of institutions, such as technology and positivist science, mean that several of the socio-economic forms traditionally held as constituting factors of society no longer serve as a valid basis for explanations. Therefore, new concepts resting on post-structuralist elements need to be found, and these concepts need to embrace the politics of signification/representation, as well as connect the immaterial to tangible practices and the material in general. Therefore, ANT, with its conceptualisation of power as vested in interaction between humans, non-humans as well as with immaterial entities, can be seen as one alternative for conceptualising post-modern society. This said, the need to anchor and organise the view that ANT provides of competing discourses, representations and actors still remains. In the following, we will discuss the particular benefits of ANT in terms of its grasp of representations of nature and the concept of power in the analysis of environmental problems like desertification, and make suggestions of possible complementary theorisations to support its explanatory/analytical capacity.

8.2.2 Actor-Network Theory and the politics of nature

The social construction of nature can be taken to mean a variety of things as pointed out by Demeritt (2001), ranging from refutation of certain claims and beliefs by revealing a tacit political intent, to drawing attention to the ontological basis of the entity. While ANT addresses the ontological understandings of nature, as well as of any other entity that might come under consideration, by showing how actors construct and assign roles to each other it also works towards revealing the political or moral claims inherent in representation and naming (see also Escobar, 1998). This has to do with the particular conception of power inherent in the ANT (see Section 8.2.3), suggesting that ANT provides a method to conceptualise the uncertainty resulting from the post-modern demise of traditional economic and institutional forms and elements of society.

Advocates of ANT support the idea that non-human resources, for example natural resources (but also, for example, technology), act to reinforce the constitution of society and, therefore, participate in actor networks, building and/or defining what society at each point in time is, just as human actors do (Latour, 1986). The ability of ANT to treat nature as a hybrid socio-natural entity distinguishes it from the structuralist and political economy approaches highlighted above, which perpetuate the destructive dichotomisation of society and nature (see also Chapter 1). This characteristic also renders ANT more capable of pursuing an *understanding* of cause-consequence relationships that underlie environmental problems such as desertification, that relate to both human and natural elements in their context. According to Callon (1986), the process of *sociological translation* shows how the interpretative frames, which we use to lend meaning to our surrounding environments, intertwine elements of the social and the natural, whether tangible or abstract, human or animal, living or inert. Taking the example of desertification, this then means that we overcome the problem created by the difficulty of pinning down an exact definition of desertification - as discussed in detail in Chapter 9. Acknowledging that action and perceptions are established and exercised through a process of *sociological translation*, and not according to some ultimate truth, not only gives weight to the *various discourses of desertification* promoting differing parameters for defining what is a problem and what not, but also shows how these interact and what implication they have for practice. This brings us to the issue of power and/or agency, and action.

8.2.3 Actor-Network Theory as a theorisation of power

ANT describes the processes through which authority and power can be attributed to certain discourses and actors, not overruling the impact of wider structural phenomena outside the immediate arenas of interaction. The notion of *agency* is, therefore, pivotal in the actor network. However, due to the hybrid nature of the network, the traditional understanding of agency - as describing the capacity to process social experience and to devise ways of coping with life within the limits of the present information and other physical, normative and politico-economic constructions (Long, 1992) - is not sufficient. Although interrelations in the network are crucial for the concept of power and agency in ANT, its capacity to incorporate non-human actors (to intertwine the social and the natural into a heterogeneous network) means that the concept of social actors is somewhat redundant. Instead, ANT rests on a Latourian understanding of power, which is vested in interaction and inconceivable without an impetus provided by an actor that will *result* in power being exercised, rather than actors' actions being explained as a causal *effect* of power (Latour, 1993).

ANT, therefore, offers means of *explaining power* by revealing the ways in which specific actors manage to impose certain outcomes on others. ANT relies on the 'actor worlds' to provide interpretations and to define and to associate the social and natural elements, which only come into a clear relationship with each other in the context of specific issues. According to Callon (1986), actors gain identity and interest through the process of translation which is integral to ANT, whereby translation is the mechanism by which the social and natural worlds progressively take form. It is a negotiation process that includes *problematisation* and, if successful, ends in *mobilisation*. Through problematisation, actors attempting to gain new authority over the issue define the problem (e.g. desertification), 'relevant' stakeholder groups, their identities, as well as the links between these different issues. The process by which a new emerging authority tries to impose and stabilise the appropriate identities and worldviews on others in the translation process is called *intressement* (Callon, 1986). Successful intressement achieves enrolment, i.e. acceptance of interrelated roles through multilateral negotiations and trials of strength. When the chain of actions is successful, there is consensus about the social and natural 'reality' presented at the problematisation stage (e.g. that 'desertification' exists as a problem), and mobilisation can then take place (e.g. policy action). Simultaneously, the defined series of hypotheses on identity, relationships and goals of the different actors are accepted and put into use, although these hypotheses are always negotiable and contestable. ANT, thus, helps to reveal these negotiation processes, struggles and redefinitions of conceptual and institutional boundaries of discourse (Ward *et al.*, 1995).

What repercussions does this discussion have on specific environmental problems such as desertification? Environmental management can be seen as a territory where the identity (what the actor is and can do) of the different actors - in our case, for example, farmers, extensionists, policies and the natural resources (e.g. soil, water) - is contested and negotiated. ANT perceives the actors as being assigned identity through interaction (i.e. the connections different actors have) in the network, hence the term 'relational materiality' (Law, 1999). What the ANT is especially useful for is what we term 'framing', i.e. disentangling the complex locality-specific networks of assigning meanings or identities. It defines individual agents, which are clearly distinct and disassociated from each other, without, however, giving an 'a priori' definition of the actor or the role of non-humans in the action. It thereby overcomes the problem of assigning specific characteristics of the network as defining characteristics from the start, and also helps focus on how, in a defined set of relations, interaction between actors recreates or contests identities and meanings (i.e. interpretations) and how and which of these become prevalent (Callon, 1999). Not only do the relations in which the entities/actors are located help define their identities, but they are also needed to express these identities which are performed in, by and through the interaction between these actors. Hence the term *performativity* is a central one to ANT, which, according to Law (1999), may also be described as 'the semiotics of materiality'.

Yet, although ANT, or the theory of translation, is in essence concerned with the mechanics of power (Law, 1992) and can help describe *how* certain types of interactions manage to stabilise and reproduce themselves, become 'macro-social' in a sense, or establish themselves as the ruling interpretation or discourse, ANT does not do much to help us understand *why* certain actors manage to impose their constructions of particular issues or places on others. This means that for causal explanations (relating to policies and desertification, for example) we need to look elsewhere, to the actual *context* of negotiations. In this context, and as suggested earlier, the *institutional setting* may provide support for certain discourses and definitions. As in political economy explanations of society, it is often the case that institutions, as with technological solutions or scientific claims, help maintain certain interpretations and representations of reality. Although the post-structuralist approaches challenge the durability of institutions, there is still a need to acknowledge their existence and role in society, while not relying on institutions as constants throughout time and space. In fact, the difference between the theory of discourse (Foucault, 1991) and ANT is one of scale - the approach the theories take to the empirical, the pragmatic. Indeed, ANT enables us to focus on the local, and to involve material, institutional and structural factors in the analysis of power on the local scale. This brings us back to our initial problem of conceptualisation. In order to derive more widely applicable benefit from the insight into the interpretative and negotiative dimensions of the 'reality' of desertification, we need to anchor our viewpoint onto some concepts that encompass the fluidity of definitions as well

as of power. In the final section of this chapter we suggest that this can be fruitfully done by combining *new institutionalist theorisations* of the policy process with the conceptualisations of power provided by ANT outlined in this section.

8.3 Combining Actor-Network Theory and new institutionalist approaches to better understand desertification

It has to be pointed out that the use of ANT is not always in rejection of other, perhaps more 'conventional', theories of rural development and environmental issues. ANT can, for example, be combined with elements of structural political economy approaches in order to add to its capacity for explanation. Murdoch (2000), for example, advocates ANT as a means to elaborate the understanding of actors, connections and spatial reach of food production provided by the *commodity chain approach*. The commodity chain approach helps to grasp the role of natural resource use in the food chain, especially the phenomena of increasing technological control of natural forces and the increasing need to transport food products over longer distances - in other words, the growing socio-technical complexity of food chains and the implications of this for the environment. In this way, it is possible to focus on the existing structures through which rural development is often constructed, and view the 'flaws' in these from the point-of-view of desired developments. These approaches particularly help investigate how the arrangement of resources affects the rural resource base. Focusing on manufacturing and services, Murdoch also focuses on spatial or 'horizontal' networks which coordinate a range of activities located within an area, thereby dealing with the capacity of actors to gain access to markets and to other economic opportunities. These networks thrive within research and development, technology, training, design, and quality policy and marketing strategies. It is, therefore, often argued that by incorporating the actor perspective to the commodity chain approach, ANT can help better explain the functioning of these learning and innovation networks.

How does this inform the approach we wish to take in Chapter 9 in our detailed analysis of actor networks operating within policy arenas in desertified areas of Southern Europe? In Chapter 9, we will apply the ANT approach to the case study material presented in Chapters 4-7, consisting of stakeholder accounts or 'narratives' of desertification, land management and farming in Southern Europe. We will move away from a prescriptive approach stipulating a pre-determined, superior way of governing, and, instead of focusing on the structural and substance-related issues of policy, we will focus on what Jones and Clark (2001) term the *modalities of governance* and the *supporting discourses* as they are portrayed in interview material (see also Chapter 3 for methodological issues of the MEDACTION Project). Thus, we argue that what makes studying stakeholder accounts meaningful from the point-of-view of trying to understand society, is the *reflexivity* of these accounts, especially as, according to Silverman (1997), they simultaneously describe and constitute social reality and the fluid structuration and interpretation of actions and institutions. The desertification narratives of the interview material, therefore, outline the networks of actors involved, and, when using the ANT approach, demonstrate how the local natural resources are involved as actors in those networks, what roles they are granted (ranging from production factors to entities that hold value *per se*), and how they act to empower other actors in the network by imposing specific interpretations over others. In combination with new institutionalist approaches outlined above, ANT, therefore, draws our attention to *process*.

Two final notes of caution are necessary before we embark on our analysis of actor networks and the implementation of policies affecting desertification in Southern Europe in Chapter 9. First, and as Adger *et al.* (2003) emphasise, it is commonly suggested that post-structuralist academic enquiry into governance should be grounded on a *thick description* of the context of governmental intervention. Acknowledging the relativist approach of ANT, this is a relatively obvious requirement that will underpin all steps of our analysis in Chapter 9. Second, and slightly more contentious, is the fact that by adopting the performative definition of power instigated by ANT, analysis should focus on mechanisms, strategies and practices, as well as on the participants of the translation process(es) under focus, including their professional affiliation and favoured discourses. As Jones and Clark (2001) rightly highlight, this also means paying attention to *variety* as well as *content*. We argue, therefore, that these are the kinds of conceptualisations that should guide academic enquiries into both environmental governance and policy evaluation. In the following chapter, we apply a comparative analysis of case study material guided by these conceptualisations.

9. Actor networks and the implementation of policies affecting desertification in Southern Europe

M. Juntti and G.A. Wilson

9.1 Introduction

This chapter provides an analysis and comparison of the effects of past policies on land managers' decision-making and desertification across the four Mediterranean case study areas described in detail in Chapters 4-7. The aim will particularly be to focus on the decision-making processes of individual actors involved in managing areas prone to desertification. It is hoped that this will enable a deeper understanding of why past policies have failed to adequately address desertification issues at the farm level, and, particularly, why in some cases policies have even *exacerbated* desertification problems. The resulting information is vital for the assessment of the role of past and present policies on desertification prevention on farms (both at national and international level), and contributes towards the improvement of existing policy mechanisms. As Chapter 8 outlined, by combining a focus on both policy-makers and the recipients of policies (i.e. the main stakeholders in the policy arena of the case study areas), this approach provides a step towards inclusive and deliberative policy-making frameworks - approaches that will assume increasing importance in the 21st century.

As outlined in Chapter 3, the collection and analysis of 'vertical' primary data was carried out by national teams of researchers working in each case study area (speaking the local languages and also well embedded in local and regional actor networks). Results of the national research in Spain, Italy, Portugal and Greece were presented in Chapters 4-7 of this book. In this chapter, we wish to focus on the 'horizontal' transnational data analysis which comprises a comparative assessment of policy impacts on desertification across the four case study areas, based on empirical data (stakeholder interviews and farm surveys) provided by the national research teams.

This chapter will adopt an actor-oriented perspective as outlined in Chapter 8, and will focus on the results of the comparative analysis of the case study data, identifying policy networks, the central policies that hold implications for desertification (see Table 2.1 in Chapter 2 which highlights the range of policies under investigation), and the processes (or driving forces) through which these impacts are induced. Methodologically, the aim of following the decision-making processes of individual land managers implies a break away from the political economy approaches to environmental management and policy implementation. As Chapter 8 emphasised, an actor-oriented approach guided particularly the comparative analysis of the interview and survey data from the four case study areas, focusing on a Latourian understanding of power as implicit in, and sustained through, interaction, rather than as something distributed as a static quality among designated actors, structures or institutions. Attention will, therefore, be drawn to the different discourses of land management and desertification, especially as the 'moral codes' or generally accepted means and ends of land management held by policy stakeholders and farmers alike hold implications for the implementation of policies governing landuse in the case study areas. Four agendas describing the perceptions and practices of policy goals and impacts will be identified from the stakeholder interviews, and their implications at farm level will be further examined through the farm survey data.

Our results not only provide a thick description of the main policies, landuse processes and institutional and cultural factors affecting desertification in the four case study areas, but also provide an overall understanding of individual policy processes and related landuse contexts and discourses that affect desertification in Southern Europe. Therefore, instead of making recommendations for new policies to mitigate desertification, this chapter will particularly highlight the cause-consequence relationships that exist when certain policies are implemented in areas with specific environmental, cultural and structural characteristics that, ultimately, perpetuate desertification. The designated pathways towards desertification mitigation involve recommendations for moderating both the processes of policy implementation and landuse decision-making and the discourses of desertification, rather than just outlining individual policies that need to be addressed (see also Chapter 10). However, we also acknowledge that the analysis of the material from the four case studies only provides a first glimpse of the complex political, economic and cultural forces at work in land management decision-making, and that it reveals the considerable challenge of uncovering and addressing the tacit barriers to environmental conservation in Southern Europe.

Building on Chapter 8, Section 9.2 will, first, discuss the complexities of understanding actor networks and power in policy implementation. The focus here will be particularly on power relationships and networks of interaction, but we will also highlight the importance of understanding discourses, agendas and central actors in the four case study areas as a basis for assessing interlinkages between policies and desertification in Southern Europe. Section 9.3 will then investigate discourses of desertification at the local level in more detail, outline five different interpretations of desertification used by various stakeholders, and discuss the importance of these interpretations for actions on the ground. This discussion will form the basis for the identification of policy agendas and desertification in Southern Europe discussed in Section 9.4. Here, we will identify three policy agendas that enhance desertification and one that may act towards desertification mitigation, and inter-dependencies between these four policy agendas will be outlined as a basis for formulating recommendations for how the policy formulation and implementation environment in Southern Europe could be improved to alleviate desertification risk. The final section (Section 9.5) then delves into four specific examples of interlinkages between policies, desertification issues, actor networks and dominant interpretations of desertification to highlight, in particular, why many of the policies currently affecting landuse in Southern Europe tend to *exacerbate* rather than *mitigate*, desertification problems.

9.2 Actor networks and power in policy implementation

9.2.1 Power relationships and networks of interaction

As the previous chapters of this book have highlighted, our analysis is a study of the effects of state intervention on desertification, and the case studies reported in Chapters 4-7, as well as the 'horizontal' collaborative analysis provided in this chapter, adopt a bottom-up theorisation of policy implementation. We argue here that giving analytical priority to the operational, grassroots or actor level of policy implementation means that the studied phenomena of desertification, the field of relevant polices, and the network of stakeholders, all become defined through the descriptions emanating from these fields of investigation, from previous research, and, most of all, from our interview and survey data. This means that desertification is addressed both as a *concept* defined by a multitude of different discourses, as well as a tangible unsustainable *process* that is connected to a complex set of policies and changing land management practices. Moreover, and following Long (1997), we argue that policy outcomes and land management practices can be viewed as outcomes of processes of negotiations between central actors, such as policy stakeholders, extensionists and advisers, scientists, local inhabitants and the land managers themselves.

By using an actor-network approach, we will, in the following sections, specifically investigate the importance of networks of actors that shape the nature and direction of policies that affect desertification processes. In this sense, the results provide a first glance into the complex policy, economic and socio-cultural contexts at local, regional and national levels in areas of Southern Europe affected by desertification, while also yielding an analysis of how these factors hinder, or promote, the implementation of policies aimed at alleviating the threat of desertification. We are, therefore, mapping the cultural and policy context in which an environmental reform of institutions and processes should take place as a result of the UNCCD policy framework (Frouws and Mol, 1997; see also Chapter 2).

The perceptions and aspirations of the policy stakeholders, acting both as observers and participants in the land management processes in the four case study areas, constituted most of our research material. As Chapter 8 highlighted, we have chosen the ANT approach (Callon, 1986; Law and Hassard, 1998) as a means to conceptualise the power relationships in the network of interaction through which policies are defined. This approach understands social relations as no longer just those existing between people, but as involving both human and non-human actors. Nature is seen to intervene in social processes, without considering natural laws as transcending those of society, but in fact as participating in the formation of, and subsequently being subjected to, social 'laws' of cause and consequence. As Castree and Braun (1998) emphasise, these *hybrid nature-cultures* can be seen as consisting of intertwining interpretative networks of nature, culture, science and technology. Chapter 8 highlighted that, through the theory of sociological translation (Law, 1992), ANT allows us to situate and contextualise the place-specific aspirations and local resources (natural, technological and financial, for example) which dictate the severity of the processes of desertification into a shared frame, in which it is possible to explain how the relevant policies are interpreted and how they condition the desertification process.

As the following discussion will highlight, such an approach reveals the organising processes guiding action, and uncovers the often unspoken moral premises of our interventions with nature, while, simultaneously, pointing out the responsible actors. This furthers the understanding of particular production processes, policy and training networks, as well as individual policies that form the tacit driving forces of desertification in our case study areas.

9.2.2 The need to understand discourses, agendas and central actors in the four case study areas

Viewed through our ANT framework, and based on the discussions presented in Chapters 4-7, it is evident from our case study data that the interaction among humans and the natural resources of the case study areas result in highly disparate policy outcomes, depending on how power is being exercised among the stakeholder networks in question. Indeed, different values and roles of involved actors result in different interpretations of policy goals which, in turn, influence policy drivers of desertification in the case study areas. However, this is not to say that structural conditions and policy design are insignificant factors, although even these are open to interpretation and acquire an instrumental role in the interview accounts of our stakeholders. This is illustrated especially by the examples of how both too much independence and lack of discretion at regional level are blamed for environmentally-detrimental policy outcomes.

In this context, it is important to highlight that other researchers have also pointed towards specific 'policy cultures and networks' that may characterise Southern European ways of dealing with environmental problems. Researchers like La Spina and Sciortino (1993) and Pridham and Konstadakopoulos (1997), for example, argue that a 'Mediterranean syndrome' can be identified (see also Chapter 2). This syndrome is characterised by structural deficiencies common to most Mediterranean countries, such as the lack of comprehensive plans or programmes to combat environmental problems and poor cooperation between the various administrative sectors that hold competence in issues such as desertification (Tsolakidis, 1998; Mourão, 1998; Pridham, 2002). Although conceptualisations of the Mediterranean syndrome have been accused of over-generalisation, symptoms of the syndrome come to the surface in our analysis. Yet, it is evident that not all blame for the problems associated with desertification in Southern Europe can be directed at the political structures and cultures of the Mediterranean member states themselves. Thus, and as Chapters 4-7 have amply demonstrated, the incompatibility of some EU policies with the natural environment that makes sustainable use of natural resources difficult also needs to be acknowledged (Beopoulos, 2002), as does the top-down nature of EU policy making (Albromeit, 1998; Wilson and Hart, 2000). Indeed, and as our Greek case study example presented in Chapter 7 particularly illustrated, it has been argued that the under-representation of regional perspectives is implicit to the decision-making processes at EU level, which may lead to an under-representation of Southern European interests at the EU policy decision-making level (Albromeit 1998; Buller *et al.*, 2000; Wilson, 2001; Beopoulos, 2002). So, not only do the structural and actor-level characteristics associated with the Mediterranean syndrome act as drivers of unsustainable land management and consequent symptoms of desertification, but also conflicting signals from Europe.

One example from our case studies may suffice to illustrate this point. The application of the EU Structural Funds and Single Market mechanisms in general is widely accepted to have led to intensification of land management practices and consequent exacerbation of environmental degradation in Southern Europe (Pridham, 2002). For example, in our Greek case study area (Island of Lesvos), desertification is linked to discrepancies between policy aims and outcomes. The LFA support has increased the number of eligible animals for the sheep and goat meat subsidy from 500 heads to 1000 per farmer (Beopoulos *et al.*, 2002). While Chapter 7 clearly highlighted that these numbers are environmentally unsustainable in the fragile pasture ecosystems of the case study area on the island of Lesvos, the LFA regulation can be seen to further support an idea of farm viability which is unsustainable for the local environment. As it is, the prefectural agricultural officials make no effort to limit the number of livestock for which farmers receive compensation from the sheep and goat meat CMO or the LFA regulation. They claim merely to be implementers of policies into practice, with no power over how this should happen in the fragmented and hierarchical administrative system comprising the Greek agricultural administration. Consequently, in Lesvos, sheep continue to degrade the vegetation and overgraze the land. As in most of our other case study areas, the most frequent source of information about subsidies for farmers in Lesvos is the cooperative - needless to say, an actor that is focused on commercial interests of farming in the area rather than environmentally sustainable herd sizes. Moreover, the Greek agri-environmental programme, of which 60% was directed towards mitigating desertification in areas such as Lesvos, remains yet to be implemented (see Chapter 7). Thus, structural drivers, in particular related to the top-down EU policy environment, have to be seen as important additional underlying forces intensifying desertification in Southern Europe.

This example shows that implementation of policies with the most significant implications for desertification have to be particularly carefully analysed using the ANT approach. Thus, specific attention will be paid to the interpretations of desertification that appear convenient for different interests and ways of managing local natural resources and that, ultimately, shape different trajectories of desertification. Different interpretative policy agendas that influence policy implementation, and that are held by different stakeholders can, therefore, be expected, and these are likely to correlate with how desertification is defined (see Section 9.3). Different agendas should also help us explain why desertification is often exacerbated by policy processes in the structurally 'weak' administrative environments of the Mediterranean countries that often suffer from a lack of attention to operative level reality in EU policy design and implementation (see Section 9.4). In this context, the following section will explore in more detail the different discourses of desertification in operation at the local level in Southern Europe.

9.3 Discourses of desertification at local level

Section 1.2 in Chapter 1 discussed the various 'scientific' definitions of desertification that are used in policy documents aiming at combating desertification. As we highlighted, a key definition is that proposed by the UNCCD that sees desertification as "land degradation in arid, semi-arid and dry sub-humid areas resulting from various factors, including climatic variations and human activities" (UN, 1994, 4). However, our data from both stakeholder interviews as well as the farm surveys yielded a number of different, arguably less 'scientific', understandings of the phenomenon. As Chapter 1 highlighted, the UN definition is sufficiently vague to allow for a heated debate between competing interpretations at global level, with implications for debates on desertification at international level as well as at the operative level, especially in small-scale localities where the UNCCD definition is often not known (Sullivan, 2000). As a result, the concept of desertification in any of its 'official' meanings is hardly recognised at grassroots level in our four case study areas (see Chapters 4-7). Yet, depending on varying political and environmental conditions, as well as types of landuses that our stakeholders are involved in, interview accounts revealed a plethora of discourses attempting to define the nature of desertification and suggested strategies to combat desertification. As Adger *et al.* (2001) argue, as most global-level definitions of desertification remain detached from the context in which associated policies and programmes are to be implemented into action, grassroots discourses help fill the gaps in understanding the cultural and context-related barriers to desertification mitigation.

The stakeholder interviews with both 'officials' and 'land managers' conducted for this research (see Chapter 3 for detail) reveal a variety of ways of interpreting what 'desertification' means. These different, more localised, interpretations of desertification are an important component of the moral codes guiding policy implementation and the use of natural resources in the four case study areas of this research, often contributing to what Lowe *et al.* (1997) and Low (2002) have referred to as *discursive barriers* to the mitigation of environmental problems such as desertification.

Perceptions of the causes and nature of desertification distinguishable in the stakeholder interviews from our four case study areas can be classified into five categories. Desertification has been interpreted by our stakeholder respondents as depopulation (we term this the 'anthropocentric' interpretation); as a water management issue (the 'reductionist agrarian' interpretation); as a climate-induced phenomenon (the 'fatalistic' interpretation); as a phenomenon caused by agricultural practices (the 'post-productivist' interpretation); and as embedded in notions of human intervention on nature (the 'holistic' interpretation) (see also Juntti and Wilson, in press). These five categories overlap to some extent, but, as will be discussed in detail in the following, each one holds a different interpretation of the role of the natural resources in the economy, the justifiable ends towards which these resources are to be used and, hence, a different morality according to which the extent and nature of desertification has been defined and is influencing how natural resources are managed[27].

---

[27] See also Lowe et al. (1997) for a similar analysis of how different moral conceptions shape the practices of pollution control in agriculture in the UK, implying variations in what is defined as 'pollution' by different stakeholders and, consequently, how thoroughly regulations are implemented by pollution inspectors at farm level.

9.3.1 The 'anthropocentric' interpretation: desertification as depopulation

In many of the Mediterranean languages, the term 'desertification' is widely used in the meaning of 'human abandonment of dwellings and land'. This was one of the most common interpretations of desertification arising from the interviews in our four case study areas and was held by about 60% of all respondents (Juntti and Wilson, in press). In the context of 'official' definitions of desertification highlighted in Chapter 1 this is particularly revealing, as the 'desertification as depopulation' interpretation is very different from the scientific, and often climate-related, definitions of desertification by official and transnational institutions. To a large extent, this highlights the perceptual gap that exists between the 'scientific' and the 'grassroots' interpretation of desertification as an environmental problem (see also Swift, 1996; Demeritt, 1996, 2001; Adger *et al.*, 2001; Eriksen, 2001).

It can be argued that understanding desertification as depopulation reflects a morality, where maintaining the rural population is seen as crucial for the welfare of the area. Different forms of land management, mainly farming and forestry, are regarded as not only economically productive functions, but also as maintainers of the natural resources of the locality - namely the productivity of the soil. Soil is seen as a factor of production, and the right of different forms of land management to the soil is inherent in this interpretation, which means that the authority of the land managers as the best guardians of the natural resources remains unchallenged. Desertification as land abandonment due to the loss of economic profitability of farming or forestry was particularly expressed by local politicians in the Italian case study area, but also by farmers in Greece, NGOs in Italy, and most of the interviewed stakeholders in Portugal. In the Spanish interviews, it was mainly agricultural actors at national and regional level and farmer organisation representatives who perceived desertification as depopulation (see Chapters 4-7).

9.3.2 The 'reductionist agrarian' interpretation: desertification as a non-farming issue

Another prominent interpretation of desertification in our case study areas can be described as the 'reductionist agrarian' interpretation that apportions the blame for desertification to non-agricultural activities and often to how common resources, like local water resources, are managed by government officials. This view was held by about 15% of interviewees across the four case study areas. In our study, this was particularly evident in the Spanish case study area (Guadalentín), where, as Chapter 4 highlighted, irrigation farming of horticultural crops has led to a flourishing of the local economy (Oñate *et al.*, 2002a). As a result, many Spanish stakeholders related desertification to the degradation of water resources, interpreted desertification largely as a technical and political problem linked to the allocation of local water resources alone, and placed the blame at bad management and administrative planning instead of blaming the farmers as the end-users of the resource. The blame is therefore allocated at the so-called secondary level of resource management, in this case the CHS (see Chapter 4) that is in charge of water management but, according to several stakeholders, neglects this task (Juntti and Wilson, in press). The CHS, in turn, claimed lack of resources as a reason for not controlling the illegal wells installed by horticultural farmers. This dismissal of farmers' responsibility is particularly blatant in the case of soil salinisation, where the depletion of aquifers has caused farmers to start irrigating with saline water, and, as a result, some areas of arable land in the Guadalentín have already become unfit for farming. Although technology to extract salt from seawater has been developed, irrigation with unsuitable water continues, as does the disregard of environmental problems caused by the saline waste resulting from these practices (Sumpsi, 2001; see also Chapter 4).

This interpretation of desertification as a water management problem has in some cases, especially in the Spanish and Portuguese case study areas, justified large-scale infrastructural developments that have redirected water from other catchments. This suggests that this interpretation of desertification is also a nation-wide interpretation of the problem, and that it is held by a very powerful group of actors. In Spain, it has to be seen as the legacy of the 'hydrological paradigm', establishing a strong tradition of treating water resources as subservient to modernisation and economic growth (Sauri and del Moral, 2001). While in Portugal it is only 'decision-making level' stakeholders that appeared to be in agreement over the benefits of these large-scale water engineering projects (see Chapter 6), in the Spanish case study area this morality appeared to be more strongly developed, and of all stakeholder groups interviewed it was only academics and environmentalists who contested and criticised the large-scale and environmentally highly questionable redirection of water from the north to the south of Spain (see Chapter 4).

The Italian case study area also offers examples of the interpretation of desertification as a non-farming, endogenous problem. As Chapter 5 highlighted, many Italian stakeholders emphasised the unrestricted right of agriculture to local natural resources and did not consider the damage caused by farming practices as particularly severe. Local politicians in Italy support the economic prosperity of the agricultural sector, as this is seen as essential for the economic development of the locality (and reflects the typically clientilistic approach in local governance). Particularly the 1992 CAP reform and the prospect of decoupling, increasingly prominent under Agenda 2000, encouraged the ploughing of environmentally valuable, previously protected, areas on even the most marginal lands. These lands had often already been abandoned due to economic policies that encouraged industrial development at the expense of the agricultural sector. However, the persistence of poor monitoring of both conservation measures and water consumption for irrigation suggests a pronounced lack of power of environmental and water authorities (see Chapter 5).

Interviews in our case study areas, therefore, often revealed a lack of acknowledgement that agriculture may be one of the key driving forces behind desertification, defending not only the position of farming as the maintainer of the natural resources of the local countryside, but also as holding the first right to using these resources. In particular, agricultural administrators and respondents from farmers' organisations often expressed the reductionist agrarian interpretation of desertification through which farming is granted a 'superior' position, justified by a significant role of agriculture in the local economy. Under this interpretation of desertification, soil, as well as water resources, are treated as mere production factors, whose sustainability is defined in terms of the needs of farming.

9.3.3 The 'fatalistic' interpretation: desertification as a climate-induced phenomenon

The interpretation of desertification as a climatically-induced phenomenon - held by 10% of respondents in the four case study areas - can be seen as the interpretation that comes closest to the 'official' global definitions of desertification highlighted in Chapter 1, but was mentioned less frequently by respondents in our four case study areas. This interpretation is, arguably, a less blatantly moralising approach than the above-mentioned interpretations of desertification. As Chapter 5 highlighted for the Italian case study area, endogenous causes of desertification were often neglected by stakeholders, which implies a relatively a-political approach, not explicitly justifying (or necessitating) any measures against the problem. Thus, contrary to the interpretations of desertification highlighted in the previous two sections, stakeholders adhering to the fatalistic interpretation do not see drought as a result of overexploitation of local water resources or linked to agricultural mismanagement, but as a consequence of climatic variation, almost independent from human intervention. Similarly, in the Greek case study area, desertification was considered by many stakeholders as a natural feature of the locality - mainly due to the predominance of erosion-prone volcanic soils - and was not seen to be related to the practice of livestock over-stocking on vulnerable hill slopes (see Chapter 7).

Although not the most common interpretation of desertification, many respondents in our four case study areas acknowledged that climate (and climate change) played an important part in causing desertification problems in their areas. However, at the same time, many of the interviewed stakeholders in all four case study areas also argued that climate-induced desertification was not high on their agenda, especially because 'not much can be done about it anyway' in terms of possible remedial actions. Particularly stakeholders who were not directly involved in land management themselves, but merely witnessed accruing problems such as drought and landslides, tended to have this fatalistic interpretation of desertification. The prevalence of this view outside of the networks of land management, in particular, may be one of the reasons for the political unattractiveness of desertification in the southern Mediterranean (and possibly at the global level), resulting in both its relative neglect as a political issue and relatively low priority in regional, national and EU decision-making agendas.

9.3.4 The 'post-productivist' interpretation: desertification caused by agricultural practices

As a fourth identifiable interpretation of desertification, about 10% of interviewed stakeholders allocated the blame for desertification to agricultural practices (Juntti and Wilson, in press). Together with the anthropocentric interpretation discussed above (desertification as land abandonment), this interpretation differs most from the 'official' definitions of desertification, which - as highlighted above - have rarely acknowledged that 'endogenous' forces such as agricultural practices can be major causes of desertification.

In Italy and Spain, several administrators, and even some representatives of agricultural interest groups, saw the problems of desertification as closely linked to over-irrigation and, more specifically, to excess expansion of irrigation farming (the reverse view of the reductionist agrarian interpretation above). On the other hand, in the Greek case study area, where the existence of desertification is widely disputed, it was mostly non-farming related stakeholders - local inhabitants and administrators - who pointed to farming practices as a cause of environmental degradation. This 'post-productivist interpretation' of desertification (see Wilson, 2001, for a detailed discussion of 'post-productivism') expresses a morality that condemns intensive agricultural practices, and that implies that both soil and water are regarded in a role that is wider than just that of a production factor in agriculture. As Chapter 7 highlighted for the Greek case, some stakeholders pointed towards the possibilities for further expanding tourism in the case study area as a long-term possible alternative to environmentally destructive farming practices in the most affected areas (Beopoulos *et al.*, 2002).

Yet, across the four case study areas, the strength with which the definition of desertification caused by agricultural practices and the ensuing moral judgements were expressed varied considerably. Some stakeholders - particularly Spanish academics and NGOs - put the case very poignantly, while many other stakeholders regarded the adverse environmental effects of agriculture as an inevitable 'externality' of modern farming (Oñate *et al.*, 2002a, 2002c). If we scrutinise the specific examples where this definition of desertification is particularly widespread, it most often appears as a key interpretation when stakeholders are associated with substantial institutional changes and restructuring, or when they are directly engaged in conservation measures. This was particularly obvious in the Italian case study area, where, especially at the regional level, even agricultural stakeholders recommended changes in relation to current water planning and use.

However, this endogenous view of desertification does not necessarily translate into improved action on the ground. Thus, currently, stakeholders expressing the post-productivist interpretation of desertification either do not assert it very strongly in policy implementation or, alternatively, they are not in a central position for implementing policies that would help alleviate the problem. As will be discussed in Section 9.4, this suggests that stakeholder perceptions of desertification are not directly correlated with actors' structural positions in stakeholder networks or in the administrative hierarchy, nor does the number of stakeholders expressing a specific interpretation of desertification appear decisive in whether this becomes implemented into action or not. As Chapter 5 emphasised, it is only in the Italian case study area that this post-productivist interpretation is beginning to be put into practice, although it continues to be impeded by administrative hierarchies. It is, however, evident that this farming-related interpretation of desertification is likely to gain ground, especially as environmental problems are becoming more pressing in the case study areas, and as this interpretation becomes increasingly institutionalised in the form of new administrative structures and discourses (Arcieri *et al.*, 2002).

9.3.5 The 'holistic' interpretation: desertification embedded in notions of human intervention with nature

The final interpretation of desertification may be termed a 'holistic' interpretation, in which desertification is understood as a wider phenomenon, and where the focus is placed on humans' negative relationship with nature (Juntti and Wilson, in press). This interpretation was only expressed by few of our respondents (about 5% of all respondents), partly because it diverges most from both the 'official' and 'popular' interpretations of desertification highlighted above, and because it is predicated on well informed and well educated stakeholder backgrounds. In this context, some stakeholders mentioned both excessive urbanisation and depopulation of rural areas as major causes for desertification, although this view was largely restricted to the academic sector. Interestingly, however, some of the stakeholders with agricultural backgrounds (e.g. members of farmers' organisation and extension officials) also occasionally suggested this interpretation.

This wider understanding of the problem considers social issues as well as natural resource management, but is expressed mainly by stakeholders, who, by not being involved at the operative level of land management, can afford this moralising approach. It is, therefore, a definition that is unlikely to be implemented into practice very often, and is in fact held by actors, who, due to their independent position (e.g. academics), lack of involvement with the practice of land management (e.g. educational sector), or due to an already explicit connection to conservation issues (some government officials; NGOs), can afford to have a wider view of the phenomenon that may even involve criticising contemporary environmentally-destructive processes.

9.3.6 The importance of these interpretations of desertification for actions on the ground

As Swift (1996) claims, and as is explicit from the above discussion, different political and bureaucratic constituencies play a significant role in shaping understandings of desertification. Several authors (e.g. Swift, 1996; Sullivan, 2000; Adger et al., 2001) discuss the various discourses of desertification at global level, mainly in the framework of the UNCCD and earlier international attempts at addressing the issue. Although such conceptualisations are largely based on empirical material from African countries, their central argument - that the global discourses differ significantly from the more contextualised understandings of desertification expressed by grassroots actors - is also clearly echoed by our findings. While global discourses have evolved according to the agendas held by central international funding and research institutions and governments (Swift, 1996; Cornet, 2002), the more localised discourses described above depict the interpretative struggles and dynamics at the operative level (Sullivan, 2000; Juntti and Wilson, in press). The resulting spectrum of views - some of which are more in accordance with 'official' definitions of desertification than others - is, therefore, an important factor impeding actions to mitigate desertification at any scale and, as will be discussed in detail in Section 9.4, also has severe repercussions for the implementation of a wide range of policies relevant to desertification.

The shared understandings of desertification and how processes of desertification are contested or addressed at local level, and especially how these translate into the adoption or non-adoption of certain policies, will be addressed in Section 9.4. On the basis of information provided at the case study level discussed in Chapters 4-7, analysing barriers for the implementation of policies that influence landuse decision-making in the four case study areas will reveal that certain policies have different implications for desertification, depending on who is in a position of power to implement (or not to implement) such policies.

9.4 Policy agendas and desertification in Southern Europe

In this section we wish to investigate what repercussions the different grassroots discourses of desertification discussed in Section 9.3 have on the formulation and implementation of policies that affect desertification processes in our four case study areas. In Section 9.4.1 we will, first, identify policy agendas that have the tendency to enhance desertification, while Section 9.4.2 will identify those agendas that may help alleviate desertification processes in Southern Europe. These policy agendas do, however, not operate entirely independently from each other. As a result, Section 9.4.3 explores the interdependencies between the different agendas and provides a conceptual model that discusses whether and how existing policy agendas can be 'shifted' to enable better management of desertification problems in the future.

9.4.1 Policy agendas that enhance desertification

The productivist policy agenda

As Chapters 4-7 have discussed in detail, there are several interpretative agendas aiming at the promotion of different interests and strategies of internalising regulatory interventions into land management in the four case study areas. We have seen that, concerning arable land, the cereal, oilseed and protein crop market organisations (e.g. EU Regulations 1765/92, 1766/92, 1251/99, 1253/99, 3072/95 and 2309/97) have tended to encourage intensification of landuse, as well as land abandonment. In some areas, this has inspired the encroachment of lands altogether unsuitable for tillage by cereal crops, while in others deep tillage has been required to help crops survive, resulting in inevitable soil degradation.

As Chapter 5 highlighted, the most pronounced examples of this come from the Italian case study area, where the unsustainable implementation of these subsidies has led to encroachment of forestlands and conservation areas (Arcieri *et al.*, 2002). A particularly problematic regulation in this respect has been the CMO for durum wheat (EU Regulations 120/67, 1765/92, 1766/92, 2309/97, 1251/99 and 1253/99 as defined in 1259/99), and the 20-year set-aside policy (EU Regulations 1094/88 and 2078/92) also appears to have had a two-fold impact in terms of desertification. While the latter has played a part in inducing extensification of landuse, it has also encouraged farmers in pursuit of maximum subsidies to plough up land that was previously left uncultivated in order to be able to claim set-aside subsidies for that land later on. Interviews from the Italian case study area highlight how farmers, aiming to maximise both set-aside and durum wheat subsidies, have caused increased degradation of soils through unsuitable tillage of the 'calanchi' and 'biancane', which are typical landforms particularly vulnerable to erosion (Phillips, C., 1998). Even the most marginal lands, abandoned in the past due to social policies that encouraged industrial development at the expense of the agricultural sector, have recently been re-sown with wheat.

Our stakeholder interviews suggested that both the traditional production-based subsidies and the increasingly prominent decoupled payments through the CAP have often been used by land managers to the maximum, aiming at increasing economic gains at the cost of sustainable farming practices. This notion of a subsidy-maximising ethos at farm level was evident among many of our respondents. Interview responses from the Upper and Middle Agri valley in our Italian case study area particularly represented this ethos, which can be termed a 'productivist policy agenda' (cf. Wilson, 2001). In this agenda, maintaining reasonable levels of economic income from agriculture is the main goal of policy implementation, and desertification is, therefore, largely policy driven (mainly by the CMOs for durum wheat and livestock as well as the 20-year set-aside measure).

As Figure 9.1 highlights, the policy outcome can be regarded as the end-result of a non-linear process of translation during policy implementation, involving several actors and interpretations. In the Italian case study area, the central actors defining the goals and practices of this implementation process are the Italian Ministry of Agriculture, farmers' organisations responsible for the extension of the policy at local level, and farmers themselves (Arcieri *et al.*, 2002). Expressing an 'anthropocentric' and a 'fatalistic' morality highlighted in Section 9.3, these actors tend to interpret desertification as a problem of depopulation or as a climate-induced phenomenon, while agricultural practices themselves are free from responsibility. As a result, both forest areas and 'badlands' are considered as production factors for agriculture and turned into arable land to secure the subsidy. Under this agenda, environmental regulations are characteristically seen as fundamentally opposed to any 'justifiable' agricultural expansion. Yet, although both conservation regulations and planning mechanisms are in place in the Italian case study area, the regional level administration, in charge of both landuse planning and the enforcement of conservation, is sidelined in the implementation network, and unable to exercise control over the use of the subsidies.

Moreover, detrimental effects of the durum wheat CMO in Italy are not only associated with policy design and implementation 'faults', but are also linked to the simultaneous non-implementation of other policies. In addition, the favourable approach to new, more productive, durum wheat varieties, resistant to different environmental conditions that had previously limited the area on which crops could be grown, is also a clear symptom of the predominant productivist policy agenda. This has encouraged agricultural intensification near urban areas, where the proximity of markets provides a comparative advantage, often coupled with more favourable growing conditions. Thus, the prevalent interpretative agenda guiding the implementation of the durum wheat CMO and the use of natural resources at local level in the Italian case study area involves a productivist approach to the implementation and promotion of the CMOs at local level, and reflects not only the farm-level subsidy maximising tendencies highlighted above, but also the fact that agriculture and agricultural income are important for stakeholders. This has led to the neglect of environmental impacts in the implementation of agricultural subsidies, enhanced by administrative sectorisation.

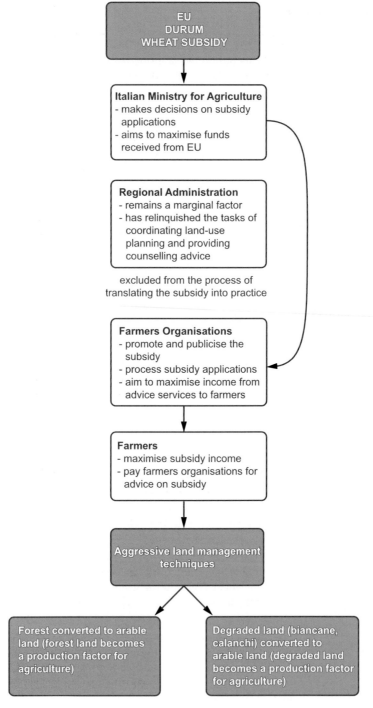

Figure 9.1: Actors involved in the implementation of the durum wheat subsidy in Italy and implications for desertification (Source: Juntti and Wilson, in press, based on Arcieri et al., 2002).

The modernist policy agenda

The 'modernist agenda', promoted in particular by national and regional-level stakeholders, encourages the development of the local and regional economy, with a focus on the most viable, technology-heavy, and, hence, 'developed' landuses. In these cases, the detrimental processes causing desertification can be seen to be more technology-driven than under the largely policy-driven productivist agenda described above. Yet, both agendas are closely interconnected (see Figure 9.2 below), and the neglect of environmental regulations remains a common feature to both agendas. According to La Spina and Sciortino (1993), this tendency to emphasise the need for economic growth and technological modernisation can be associated with the typically late development of industrialisation in Mediterranean countries.

One example of the modernist agenda comes from our Portuguese case study area (see Chapter 6). Portuguese membership of the EU in 1986 considerably enhanced the process of modernisation of the arable sector, with money from funds directed at farm modernisation for investments into new machinery (Eden, 2002). While cultivation methods intensified and became mechanised, arable land started encroaching on forested areas. Although, as Chapter 6 emphasised, high state-subsidised cereal prices had already encouraged the spread of arable farming, mechanisation induced significant changes to the biodiversity-rich *montado* landuse system, which has traditionally combined livestock, forest and cultivated areas in environmentally sustainable ways (Vieira and Eden, 2002). The reduction in tree numbers is, therefore, a direct result of the mechanisation of arable farming, as maintaining a combination of trees and arable crops became increasingly difficult with the advent of larger machinery. This was amplified by EU cereal subsidies, which (as in the Italian case study area described in Chapter 5) encouraged the spread of arable farming onto *montado* lands that was not entirely suitable for this use (Roxo and Casimiro, 1998; Vieira and Eden, 2002). This not only led to an increase in expenditure of irrigation water, but also caused deterioration of the soil. However, Chapter 6 highlighted that Portuguese stakeholders mainly interpreted desertification as depopulation of the countryside, and this 'anthropocentric interpretation' can be seen as one of the key factors impeding both the recognition of these symptoms and the willingness to implement policies that may mitigate them. Yet, while these changes took place in the wake of Portuguese EU membership in the 1980s, recent trends may suggest a possible reversal of this process, with an increase in farmers' awareness of erosion and the loss of trees on the *montado*.

The modernist agenda is perhaps strongest in the Guadalentín valley in the Spanish case study area where we witness distinct neglect of environmental regulations, uncontrolled spread of irrigation farming, and the marginalisation of dryland farming (see Chapter 4). Maintenance of the latter would, in particular, allow the continuation of relatively sustainable forms of landuse on hill-slopes surrounding the intensively irrigated farming areas in the valley (Oñate and Peco, 2005). The approach of stakeholders towards policies concerning the quality and use of water serves as a case in point, as these policies have, so far, proven ineffective in terms of regulative targets, mainly due to non-implementation. Consequently, water extraction for intensive farming remains an economically feasible and accepted practice, even in an adverse policy environment. Thus, in the Guadalentín valley the spread of irrigation farming is symptomatic of the modernist agenda, largely driven by technological development and by a buoyant market for horticultural products, the non-implementation of regulative policies, as well as particular interpretations of EU farm modernisation aids outlined in Chapter 4. Contrary to the Portuguese case study, in the Guadalentín area farm modernisation aids have been used to expand irrigated land at the expense of dryland farming. Even the direction of research and the development of technology are all geared towards serving the purposes of irrigation expansion and efficiency. As a Spanish national level environment official (National Water Authority/Ministry of Environment) argued, "agriculture is no longer linked to soil, which acts just as a physical support" (Oñate *et al.* 2002a, 35).

As a result, in the Spanish case study the modernist agenda is strongly embedded in the above-mentioned 'reductionist agrarian' interpretation that sees desertification solely as a problem of water management and allocation. While in the Guadalentín this interpretation was promoted mainly by farmers (and, to some extent, also by the powerful cooperatives they belong to) and the regional agricultural administration, the modernist agenda became institutionalised in the Spanish *National Hydrological Plan* passed in 2002 (which transfers water from the Ebro basin to the south of Spain; see Chapter 4). Although the formal aim is not irrigation expansion but protection of aquifers in southern Spain, several interviewed respondents implied that the actual aim is continuing irrigation expansion. The environmental benefits of the plan depend on the efficiency of monitoring and control undertaken by both the regional branch of the CHS (the national water authority) and the local/regional agricultural administration. However, the agricultural administration is the same body that backs irrigation expansion under the modernist agenda, while the CHS, so far, has exercised only marginal influence to enhance sustainable management of water resources. It appears, therefore, that while farmers, cooperatives and the regional agricultural administration are the central proponents of the modernist agenda, the water administration is - either voluntarily or through lack of power in the stakeholder networks - forced to lend its support to the same goals.

Irrigation expansion in Spain and the lower Agri in Italy, where irrigated horticulture has recently dramatically increased, as well as the spread of arable farming in Portugal, also appear to be a result of a more or less determined attempt on the part of the regional-level administration (mainly through agricultural officials, local politicians and farmers' organisations) to enhance the technological and economic development of the regions (Juntti and Wilson, in press). In addition to intensification and expansion of cultivation onto vulnerable soils, this often involves neglecting legislation that restricts the use of vulnerable ecosystems for production. In both cases, stakeholder interpretations of desertification that denounce the responsibility of land management to solve large-scale environmental problems underlie these developments. In particular, the above-mentioned 'anthropocentric', 'agrarian reductionist' and 'fatalistic' interpretations of desertification emphasise economic gains to the locality and the furthering of development as causes that justify the lack of attention to environmental impacts. As a result, in Spain the regulative measures aiming at protection and sustainable use of water resources and irrigation land (see Chapter 4 for details) have become subordinate to farming, evidence for which is provided by illegal water extractions and sale of water concessions, and the continued and illegal rotation of irrigated crops over the years.

The modernist policy agenda can, therefore, not only be identified through recent historical landuse trajectories in Spain, but also through the way in which any regulative policies focusing on the use of natural resources are implemented or not. Characteristic for the Mediterranean syndrome, local authorities are reluctant to enforce regulations that are seen as contradictory to the interests of their 'clients' (Briassoulis, in press). The 'animus regandi' (Oñate *et al.*, 2002a) is one of technology-driven economic growth achieved by irrigation farming, and any initiatives that might threaten these aims are regarded as damaging to the development of the region. As Chapter 5 emphasised, such tendencies are also visible in the interviews of the Italian stakeholders. While administrative actors in the lower Agri recognised the problem of desertification resulting from hydro-geological instability, their concerns were not reflected in actions in policy implementation 'on the ground'. This suggests that not only structural conditions, but also both the modernist and productivist agendas - nurtured among central policy stakeholders on the basis of relatively 'narrow' interpretations of what desertification means and entails - are to blame.

The instrumental policy agenda

This policy agenda is characterised by policy stakeholders interpreting desertification as an anthropocentric problem, mainly as 'land abandonment' and 'depopulation'. Whether this reflects the actual nature of the problem - land degradation having advanced to a point where arable land is abandoned by farmers incapable of producing a sustainable yield - or not, this agenda holds specific implications for the implementation of funds directed at land conservation (Juntti and Wilson, in press). In this context, we focus our attention on the role of EU Agri-Environmental Policy (AEP). Whereas EU Agri-environmental Regulation 2078/92 combines both farm income and environmental aims (Buller *et al.*, 2000), it is the role of agriculture in maintaining both social and environmental dimensions of the countryside that is emphasised by stakeholders who adhere to an instrumental policy agenda. Thus, interpreting desertification as depopulation, with associated concerns over the abandonment of agricultural land due to decreases in farm profitability, appears to inspire this instrumental approach to farming, forestry and conservation subsidies, with the result that AEP subsidies are considered more as income support than conservation measures (see also Juntti and Potter, 2002).

As our analysis in Chapter 4 highlighted, the instrumental policy agenda is particularly evident in the hill farming areas of the Spanish case study area, where agricultural abandonment, along with unsustainable cereal and tree crops encouraged by available EU subsidies has been a problem (see Oñate *et al.*, 2002a). In particular, stakeholders with close connections to farming interests perceive AEP as instrumental in both maintaining farming and ensuring subsidy income. Bruckmeier and Patricio (2002) also cite both human and physical desertification as the main agri-environmental problem identified by the Portuguese implementers of the Regulation, and claim that available agri-environmental measures are mainly aimed at maintaining farming in economically marginal areas - claims also supported by results from our Portuguese stakeholder interviews (see Chapter 6; see also Vieira and Eden, 2002).

As the discussion in Chapter 5 amply demonstrated, the instrumental policy agenda can also be identified throughout the Italian case study area, where land reclamation and transformation of vulnerable lands has, since the 1950s, been partly financed by funds through the CASMEZ, directed at the regional development of underdeveloped rural areas in southern Italy (Arcieri *et al.*, 2002). From the 1950s to the 1970s, the CASMEZ was a national institution that was transferred to the regional level in the 1980s. While the combination of economic and environmental sustainability was relatively successfully undertaken in earlier projects, in the late 1960s the tone of intervention changed. As Chapter 5 illustrated, CASMEZ funds began to be seen as a means to maintain employment and economic incomes in the area. Afforestation measures financed from the funds in the 1960s and 1970s serve as a case in point. Whereas in the 1950s the coupling of employment and natural forest management served a justifiable purpose, in the 1960s and 1970s the aim of ensuring employment opportunities for forest workers took over as the main goal at the expense of environmental sustainability. Aiming for funds to provide employment for a maximum number of forest workers, the planted species were chosen according to market value of wood or subsidy levels and, therefore, often failed to survive in marginal areas. Further, the natural vegetation, such as the *maquis* of the drier lands, could not be maintained due to lack of eligibility for afforestation funds. Although maintaining forest cover would serve an important stabilising purpose against desertification, the introduced species and the lack of maintenance of the planted woodlands have severely diminished the success of afforestation measures in helping alleviate desertification.

The instrumental policy agenda, therefore, has particular implications in terms of the outcome of conservation regulations. Although some results are beginning to be achieved through AEP, particularly with regard to regeneration of wooded areas across Southern Europe, several stakeholders see AEP payments as insufficient to achieve significant improvements (see also Wilson *et al.*, 1999; Buller *et al.*, 2000). Further, although take-up of available AEPs has been high, Chapter 4 emphasised that in the Spanish case study area relatively little evidence of actual environmental benefits exists, and monitoring, as well as provision of advice and promotion, appear to be lacking (Peco *et al.*, 2000). Oñate *et al.* (2002a), therefore, suggest that the administrative bodies have focused on supporting irrigation farming, and that these bodies have tended to pay little attention to areas of less productive marginal farming. This could be a factor enhancing the instrumental policy approach vis-à-vis available AEPs, most of which are not applicable to intensive irrigation farming. Thus, in the Spanish case study area, the modernist policy agenda reinforces the instrumental agenda with regard to the relatively weak implementation of desertification mitigation measures. This bears resemblance to the 'distorted incorporation' of EU environmental legislation at national and local levels ascribed to the Mediterranean syndrome by La Spina and Sciortino (1993), and is in line with what Bruckmeier and Patricio (2002, 58) claim to be "an institutionalised contradiction between mainstream modernisation and ecological adaptation" hindering the successful implementation of AEP in Spain.

9.4.2 Policy agendas that act towards desertification mitigation

In the previous section, we have seen that the three policy agendas closely related to what others have termed the 'Mediterranean syndrome' provide a powerful explanation for specific policy preferences, interpretations and implementation failures in our four case study areas. However, conventional theorisations of the Mediterranean syndrome have received criticism ranging from denial to accusations of over-generalisation between countries that exhibit varied administrative and policy contexts (Börzel, 2000). Particularly Kousis (2000) points to an increased agency in civil societies of southern member states and a move away from the 'weak civil society model' referred to by, for example, La Spina and Sciortino (1993). Critics suggest that there is increasing evidence of environmental interest groups and protests in Southern Europe, particularly at local level, which, according to Kousis (2000), has increased the effective implementation of specific policies that also contain conservation-oriented goals. The interview data from our four case study areas partly supports this criticism, and indicates that environmental concerns related to desertification are, to some extent at least, acknowledged in the networks of stakeholders.

In this context, what we term the *reactive conservationist policy agenda* (see Juntti and Wilson, in press) can be seen to be associated with desertification-mitigating behaviour and more explicit conservation aims. A classic example of how conservationist goals emerge as a result of obvious detrimental changes in the state of the environment can be identified in the lower Agri in the Italian case study area, where, in the case of irrigation farming, a combination of the productivist and modernist policy agendas have led to excessive exploitation of water for irrigation (see Chapter 5). Since the 1990s, a series of droughts, changes in microclimate, and pollution of groundwater by seawater, have all led to a re-evaluation of water management. The neutral 'fatalistic' interpretation of desertification as a climate-induced problem and the 'agrarian reductionist' discourse linked to water allocation have partly given way to interpretations that allocate the blame more explicitly to human activities. A tangible consequence of this is that the Italian environmental administration has started to receive increasing power over water and soil issues (Arcieri *et al.*, 2002), and even the farming-oriented stakeholders are beginning to acknowledge the necessity to control water use in the region. The regional administration has started providing economic incentives and technical assistance and advice for better planning and management of irrigation, and has established a body to control water consumption (the Integrated Water Service). Arguably, the 'post-productivist' interpretation of desertification is gaining ground, with associated institutional changes and power shifts. Although desertification continues to be treated mainly as a problem of water management, agricultural interests are no longer allowed to predominate in the use of local resources.

It appears, therefore, that in the southern parts of the Italian case study area, due to increasing desertification problems, both structural problems linked to the Mediterranean syndrome (mainly administrative sectorisation) and the modernist policy agenda may begin to give way to an ethos of sustainable water resource management. In Portugal, meanwhile, although we found no indication of institutional changes, the instrumental agenda also appears to be supportive of both the implementation of AEPs and the withdrawal of direct subsidies for arable crops. This, together with increasing concern for erosion and the loss of trees on the cultivated areas of the *montados*, also suggests the emergence of a reactive conservationist agenda, pushed by changes in EU subsidies, and leading to the gradual disappearance of the modernist agenda of the 1980s and the 1990s (see Chapter 6). From a point-of-view of achieving lasting changes involving both structures and attitudes, the Italian and Portuguese cases, therefore, appear most promising.

In the Guadalentín valley in Spain, on the other hand, stakeholders have reacted to the pollution of water resources from agricultural sources by opting for short-term concealment of the problem, rather than by addressing underlying reasons. As Chapter 4 emphasised, instead of enhancing the position of the environmental authorities and coordinating actions of the CHS, several stakeholders considered the treatment of agricultural effluents as a solution to the current problems. This 'end-of-pipe' strategy has traditionally been considered less sustainable than solutions that address structural and attitudinal reasons for unsustainable resource management (Wilson and Bryant, 1997), particularly as it does not challenge the root causes of the problem at hand and, hence, does not condemn the activity. Yet, the situation can be described as slightly more promising in the more extensively used hill-slope areas in the Spanish case study area, where the agri-environmental measures are likely to start gaining ground as an alternative to other declining subsidies.

The Greek interview results, meanwhile, suggest a situation lacking dynamic forces of change (see Chapter 7). There, it is difficult to see how the gridlock following administrative sectorisation and rigid hierarchies would loosen its grip without deliberate attempts of the higher-level administration to decentralise decision-making in policy implementation. This is exacerbated by the lack of competing interpretations of desertification (see Section 9.3) and a virtually non-existent critical engagement of the local population with desertification issues in the case study area.

9.4.3 Inter-dependencies between the four policy agendas

Although we have treated the four policy agendas above as separate conceptual entities, our discussion has highlighted that these agendas are not mutually exclusive or target area specific. They should, therefore, be understood as broad conceptualisations of the social logic of landuse and implementation of related policies, identifiable through interview responses and 'desertification story-lines' of individuals involved in policy implementation. In particular, and as Figure 9.2 highlights, the three policy agendas that are related to enhancing desertification have to be seen as closely interrelated. Thus, the productivist and modernising agendas hold specific implications for the implementation of the CMOs for cereals and livestock and the farm modernising subsidies, and justify the non-implementation of environmental regulations. The influence of these agendas is reflected in the implementation of funds earmarked for conservation, leading to an instrumental approach to such subsidies and a likely dilution of environmental benefits. As discussed above, this was particularly evident from the application of the CASMEZ afforestation funds in Italy, which can be seen to reflect both the modernist and productivist agendas prevalent in land management at the regional level. While the Italian case study area comprises the most diverse selection of landuses and policy agendas, interconnections between the productivist and instrumental agendas are also apparent in the Guadalentín in Spain, where

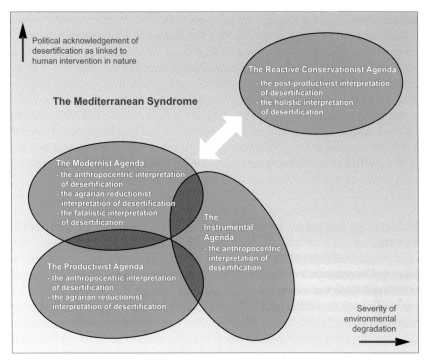

Figure 9.2: Interdependencies between the four policy agendas of desertification in Southern Europe (Source: Juntti and Wilson, in press).

both CMO subsidies and market prices encourage a shift from cereal production to the more detrimental almond and olive crops, and where AEP is perceived as a mere continuation of these subsidies.

As Figure 9.2 suggests, the modernist, productivist and instrumental agendas represent perceptions and actions within the structural conditions conceptualised as the Mediterranean syndrome, and explain the tendency of land management policies to enhance desertification. However, the reactive conservationist policy agenda depicts a new emerging interpretation of desertification that may encourage institutional changes and the formulation and adoption of policies that will, ultimately, help mitigate desertification. The conservationist agenda, therefore, has to be seen as a competing and still relatively uncommon interpretation of the goals and practices of landuse in Southern Europe.

As the concluding Chapter 10 will further emphasise, it is in the competing territory between the three 'detrimental' agendas and the 'conservationist' agenda that both the most severe *tensions* and potential *solutions* with regard to coordinated efforts for addressing desertification in Southern Europe will be found in the future. Only once these tensions between competing interpretations and policy agendas can be overcome, will effective and successful implementation of policies addressing desertification be possible. Despite the positive developments described in the Italian stakeholder interviews (see Chapter 5), institutional changes and admission of human responsibility for the degradation of soil and water resources are currently only occurring in reaction to considerable deterioration of natural resources. This can be read as a characteristic of the persistent, albeit weakening, Mediterranean syndrome. As Figure 9.2 shows, our results suggest that the shift towards a more conservationist policy agenda in the distant future will probably require further deterioration of natural resources in the near future, where desertification will be directly felt by the general public (e.g. through complete loss of soil productivity in some areas). This, in turn, may eventually lead to a recognition of the human causes of desertification by all stakeholder groups, and to a substantial modification of interpretations of the key drivers of desertification that will also acknowledge the importance of human agency and the policy environment as key triggers of desertification problems.

9.5 Interfaces between policies and desertification in Southern Europe

The discussion in Chapters 4-7, together with the analysis of actor networks, interpretations of desertification and the identification of different policy agendas in Sections 9.3 and 9.4, allow us to interpret in more detail the *interfaces* between policies and desertification in our four case study areas. In this final section, we wish to highlight these interfaces by selecting four key examples from our case study areas. Reflecting our discussions in this book so far, the focus will be on the relevant policy processes driving specific activities, the underlying driving forces for desertification linked to these policies, the central actors involved in often detrimental landuse decision-making processes, the marginal actors or 'silent voices' often sidelined in the main landuse decisions, and the dominant interpretations of desertification that provide the discursive framework within which these actions (and non-actions) are embedded.

Table 9.1 shows the example of irrigation farming of horticultural crops, prominent in parts of the Guadalentín Valley in Spain, the lower Agri Valley in Italy and in the Alentejo in Portugal. As Chapters 4-6 have highlighted, the key policy processes that affect desertification in these areas are linked to EU and national investment aids (especially through farm subsidies) and to the non-implementation of environmental policies. The resulting driving force for desertification has been expansion of areas of intensive irrigation farming, further exacerbated by more frequent occurrences of illegal water abstractions, all leading to increasing threats of salinisation and further degradation of fragile soils in these areas. We have seen that the central actors involved in these environmentally-detrimental landuse decision-making processes have been commercial farmers, cooperatives and farmers' organisations in particular, aided by agricultural officials, local government and parastatal agencies. However, as our discussion above has highlighted, these actors have only assumed powerful positions in the local actor networks because other actors have allowed them to do so. In our example of irrigation farming and horticultural crops in Spain, Italy and Portugal, these 'silent actors' have particularly included water and planning authorities (i.e. part of the regional administration), who have not been able to challenge the reductivist and anthropocentric interpretations of desertification that dominate stakeholder discourses in these areas.

Table 9.1: Interfaces between policies and desertification: the case of irrigation farming and horticultural crops (Source: authors).

Irrigation farming of horticultural crops: Guadalentin Valley (lowlands), lower Agri Valley, parts of Alentejo

Key policy process(es)	Driving force of desertification	Central actors	Marginal actors	Dominant interpretations of desertification
Investment aids (EU and national)	Expansion of area of intensive irrigation farming	Commercial farmers, cooperatives, farmers' organisations, agricultural officials, local government, parastatal agencies	Water authorities, planning authorities (regional administration)	Reductivist, anthropocentric
Non-implementation of environmental policies	Illegal water abstractions	Agricultural officials, local government, parastatal agencies	Water and planning authorities	Reductivist, anthropocentric

Our second example (Table 9.2) concerns the expansion of durum wheat monocultures that have become a particularly important driver for exacerbating desertification processes in the middle Agri Valley in Italy (see Chapter 5) and the Alentejo in Portugal (see Chapter 6). Here, the key policy processes involved have been the durum wheat CMO in particular, and the 20-year set-aside policy linked to the EU Agri-environment Regulation to a lesser extent. These policies have encouraged farmers in the middle Agri and Alentejo to convert land unsuitable for cereal production to arable land, combined with encouraging unsuitable cultivation methods, with concurrent increases in desertification and land degradation processes. Protective tree cover has also often been removed in the process - a situation particularly detrimental in the traditionally farmed and environmentally relatively sustainable *montados* of the Alentejo (see Chapter 6). The 20-year set-aside policy is a particularly interesting example here as, contrary to common belief and research results from elsewhere in the EU (cf. Buller *et al.*, 2000), it has encouraged further intensification of farming in our Italian and Portuguese case study areas, by encouraging farmers to plant cereal crops in unsuitable areas before 'setting-aside' the same areas and claiming the subsidies. Central actors in these processes have been small and/or part-time farmers and farmers' organisations, supported directly or indirectly by regional-level agricultural officials, while planning authorities (as part of the regional administration), as well as conservation/environmental authorities, have, once again, been sidelined in the actor networks as their more conservation-oriented interpretations of desertification do not fit in with the dominant fatalistic and anthropocentric interpretations that appear to guide most actions in the local actor networks.

Table 9.2: Interfaces between policies and desertification: the case of durum wheat monocultures (Source: authors).

Durum wheat monocultures (middle Agri Valley, Alentejo)

Key policy process(es)	Diving force of desertification	Central actors	Marginal actors	Dominant interpretations of desertification
Durum wheat CMO	Conversion of unsuitable land to arable land; removal of protective tree cover (especially in Portuguese *montados*)	Farmers, farmers' organisations, regional-level agricultural officials	Planning authorities (regional administration), conservation/environmental authorities	Fatalistic, anthropocentric
20-year set-aside (EU Agri-environment Regulation)	Conversion of unsuitable land to arable land; encouragement of unsuitable cultivation methods	Small and/or part-time farmers, farmers' organisations, agricultural officials	Environmental authorities	Fatalistic, anthropocentric

A third theme of policy and desertification interfaces that permeates our discussions in this book is based on policy drivers encouraging more intensive stocking of pastures with sheep and goats (Table 9.3) - processes particularly prominent in the Island of Lesvos in Greece (see Chapter 7) and the Alentejo in Portugal (see Chapter 6). The key policy processes involved here have particularly been the sheep and goat meat CMOs, LFA regulations, subsidies for young farmers, as well as the EU Agri-environmental Regulation and cross-compliance. The former three policies have particularly encouraged overstocking on steep pastures on the Island of Lesvos and in the montados, with resultant negative effects with regard to desertification processes. In accordance with other research on the impacts of EU AEP (Buller at al., 2000; Wilson and Hart, 2000), the Agri-Environment Regulation, meanwhile, has partly helped mitigate desertification problems by encouraging extensification of stocking rates in compliance with nationally-defined 'codes of good agricultural practice' (see Chapters 6 and 7). It appears, however, that the above-mentioned livestock subsidies have played a greater role in influencing farmers' decision-making, leading to increased stocking densities on some pasture areas, and thereby worsening desertification processes in already environmentally-fragile environments. Central actors involved in these processes in Greece and Portugal have been small marginal farmers, cooperatives and municipal officials who have all adopted anthropocentric interpretations of desertification, thereby sidelining agronomists and members of the regional administration who often only see themselves as mere implementers of policies emanating

Table 9.3: Interfaces between policies and desertification: the case of sheep and goat farming (Source: authors).

Intensification of sheep and goat farming (Island of Lesvos, Alentejo)

Key policy process(es)	Driving/mitigating force of desertification	Central actors	Marginal actors	Dominant interpretations of desertification
The sheep and goat meat CMOs; LFA regulations; subsidies for young farmers	Overstocking on steep pastures on Lesvos (Greece) and in the montados (Portugal)	Small marginal farmers, cooperatives, municipal officials	Agronomist, prefectural directorate (regional administration)	Anthropocentric
The Agri-environmental Regulation and cross-compliance	Extensification of stocking rates in compliance with code of good practice	Small mixed farmers	No information	Post-productivist, fatalistic

'from the top' (see especially Chapter 7). Nonetheless, and as Section 9.3 has emphasised, stakeholders involved in implementing the Agri-environmental Regulation of the EU have also occasionally expressed a post-productivist interpretation of desertification that at least acknowledges the detrimental effects of agriculture on desertification processes in these areas, although many continue to show a more 'typical' fatalistic attitude that apportions blame to 'external' factors such as climate change (see Section 9.3). As a result, compliance with the Agri-environmental Regulation and reduction of stock size does not always reflect the acknowledgement of a link between farming practices and desertification.

Our final example looks more closely at olive and almond plantations in upland areas of the Guadalentín (Spain) and Alentejo (Portugal) (see Chapters 4 and 7) - processes that, on the whole, have exacerbated desertification processes. The key policy processes involved here include, in particular, the dry fruits and almonds CMO and the olive CMO which both have encouraged farmers to plant tree crops in upland areas of the Guadalentín (moderate exacerbation of desertification risk), as well as leading to shifts towards more intensive forms of olive cropping in the Alentejo (more significant increase in desertification risk). Central actors involved here have been mainly small marginal farmers, who, as highlighted in Section 9.4, appear to be able to exert significant power in local actor networks (as they are not challenged in their views by other stakeholders) and who adhere, largely, to anthropocentric interpretations of desertification.

In conclusion, the discussion in this chapter has served to highlight two key points. First, we can only fully understand the interlinkages between policies and desertification by

Table 9.4: Interfaces between policies and desertification: the case of olive and almond plantations (Source: authors).

Olive and almond plantations in upland areas (Guadalentin, Alentejo)

Key policy process(es)	Driving/mitigating force of desertification	Central actors	Marginal actors	Dominant interpretations of desertification
Dry fruits and almonds CMO	Encouragement of tree crops in upland areas in the Guadalentin	Small marginal farmers, cooperatives	?	Anthropocentric
Olive CMO	Shift to more intensive forms of olive cropping in the Alentejo	Small marginal farmers	?	Anthropocentric

understanding both *actor networks* in operation in our case study areas and the differing *interpretations of desertification* that individual stakeholders or stakeholder groups hold. This allows us to understand why different policy agendas exist in the case study areas and, most importantly, why there can be no single solution to combating desertification in Southern Europe. Second, the results presented in Tables 9.1-9.4 also support our original hypothesis (see Chapter 2) that most of the *agricultural subsidies* available to farmers have tended to lead to further *intensification* of both arable and livestock farming in our case study areas - despite so-called recent 'reform' of the CAP through Agenda 2000 towards more environmentally-conscious policy-making. Our results have particularly highlighted that some policies that have so far been largely perceived to be benign with regard to desertification, such as 20-year set-aside policies, have, in fact, partly *exacerbated desertification processes* by providing inducements to farmers to plough up hitherto more extensively used areas. However, these 'negative' policy effects have only been allowed to unravel due to specific power structures in existence in actor networks in our case study areas - networks that largely exclude critical voices that may challenge the productivist outlooks that tend to dominate action and thought in these areas. Our discussion throughout this chapter has, therefore, highlighted that it is important to *unravel power positions within actor networks of desertification* in Southern Europe, and to also make room for understanding the 'silent voices' and 'voids' that currently exist in policy implementation hierarchies - silences that currently severely affect desertification mitigation in our case study areas.

10. Conclusions: desertification in Southern Europe: the way forward?

G.A. Wilson and M. Juntti

In this final chapter we wish to draw together the various strands of our analysis. Throughout this book, we have shown that 'unravelling' desertification involves not only understanding the physical processes that lie behind desertification problems in Southern Europe, but that a *thickly textured social science analysis* of policies, politics, power, actors and attitudes is necessary to understand why different stakeholder groups approach the 'problem of desertification' in different ways, and why no unanimous solution has been, or indeed can be, found.

10.1 The global and international dimensions

There is no doubt that both understanding and combating desertification assume ever greater importance in the 21st century, and that desertification is now firmly embedded on the global agenda. This is possibly best highlighted by the award of the *Nobel Peace Price* in October 2004 to Wangari Maathai, Kenya's Deputy Environment Minister, for the planting of 30 million trees against erosion and desertification in Kenya as part of Maathai's innovative 'Green Belt Movement'. Not only does this show that combating desertification is now recognised as one of the key 'rewardable' pursuits in the international environmental community, but, in a politically seemingly increasingly unstable world threatened by terrorism and unjustified wars, it also highlights the important link between desertification and environmental *security* - issues that will assume ever greater importance in an increasingly overcrowded world in the following decades. In this context, the actor-oriented approach adopted in our analysis of policy effects on desertification in four case study areas in Southern Europe, together with our focus on perceptions and interpretative power rather than structures, policies and their tangible outputs, allows us to draw conclusions that address the *politics*, rather than merely the *policies and processes*, of desertification.

Since EU policies form the main framework for governmental intervention into land management in our four case study areas (as well as in other Member States), the most important point to be made on the basis of this research into the links between policies and desertification is that many of the criticisms already aimed at EU policy-making hold true. First, it is blatantly obvious that the rigid top-down policy model of the EU and the lack of both accountability and consultation at regional and local level play a key role in creating the environmentally-unsustainable policy outcomes described in Chapters 4-9 of this book. Second, our results also show that these 'negative' policy outcomes are exacerbated by the fact that the UNCCD policy arena only provides a *framework* for action, and that there is a lack of *regulatory* policy drivers for change emanating from the UNCCD framework. In this context, the effectiveness of the newly established NAPs for combating desertification -

outlined in detail in Chapters 4-7 - remains yet to be proven. Third, the effectiveness of existing policies has been further hampered by the different policy agendas and interpretations of desertification held by various actors in the policy implementation networks in Southern Europe. Our results from Chapter 9 have shown how vital it is to understand the role of stakeholders in policy implementation structures, to identify the different discourses of desertification that influence whether and how 'desertification' is perceived as a problem and how it is to be solved, and to recognise that any drafting of new policies affecting localities in Southern Europe characterised by specific desertification-related problems needs to include local stakeholders in all stages of the policy continuum from policy formulation, design, implementation and, arguably most importantly, monitoring of policy success or failure.

Based on the results presented in this book, we can identify six specific reasons why detrimental and ineffective policy outcomes have predominated in our case study areas in Spain, Italy, Portugal and Greece:

1. Agricultural actors tend to dominate decision-making about using (and abusing) subsidies at farm level, and both environmental expertise and interests are largely missing from central actor networks in all our case study areas. As a result, natural resources like soil and water are often perceived as production factors and sources of farm-level and local economic benefit, rather than as natural resources that should be managed sustainably to alleviate desertification.
2. Attempts to maximise income accrued from EU subsidies in the local area, the use of both EU and national investment aids to promote intensive and technologically-advanced production (the productivist and modernising agendas respectively), and the dominance of agricultural interests in the policy networks, all increase the degradation of both soil and water resources.
3. Erosion damage could have been prevented, had the already existing conservation areas and measures been enforced more rigidly. This means that with regard to monitoring conservation efforts, particular attention has to be paid to the goals and interests enforced by actors involved in policy implementation, especially in areas where the modernist and productivist agendas dominate that have both inspired an instrumental approach to conservation policies.
4. Pockets exist where changes towards environmentally sustainable application of subsidies and investment aids are emerging (the reactive conservationist agenda), but these changes are mainly a response to an already severely degraded environment (and, to a lesser extent, to EU environmental policies) rather than being engendered by endemic forces from 'within'.
5. Cultural factors and the structure of landownership have an effect on environmental sustainability and policy outcome. For example, increasing land prices (partly due to EU subsidies) constitute a common problem in all case study areas and, together with a traditional attachment to property and legislation which favours smallholding, impede the shift towards bigger, full-time farming units. Indeed, both arable and agri-environmental subsidies can encourage part-time and absentee farming and a productivist agenda where subsidy income is maximised through expansion of arable areas without regard for sustainability. However, small farm size and attachment to land is not necessarily a cause of desertification, and evidence from Chapters 4-7 showed that

owner-occupied land is often farmed more sustainably than rented land, and small mixed farm units often exhibit more sustainable uses of EU investment aids than larger holdings.

6. Possibly the most important result of our study has been the recognition that the way in which desertification is understood and addressed is *politically charged* and linked to the prevailing consensus over the 'best' and 'most justified' use of the local natural resources. These are often manifested in, and institutionalised through, the allocation of funding towards large-scale infrastructural developments, for example transfers of water for irrigation from adjacent catchments.

10.2 Recommendations for future desertification mitigation in Southern Europe

In conclusion, we wish to outline recommendations for processes and policies that deserve further attention in desertification mitigation and that may give an idea of the kind of issues which deserve further attention in policy implementation and evaluation - especially in the context of future work following the insights gained from interpretative, actor-oriented approaches in policy and desertification analysis presented in this book.

With regard to *administration of policies and landuse planning*, our results have shown that the location of power in the actor networks of desertification requires attention, so that a better balance between agricultural and environmental interests and expertise can be established. This means not only that policy implementation needs to be more sensitive to the cultural and environmental context of land management but that institutional changes and policy learning also need to take place. This also has repercussions for *public participation*, and the administrative processes of planning and policy implementation need to be made more transparent. In particular, channels for public participation need to be established according to the requirements set in the NAPs and in line with the principles of better policy and issue ownership. This clearly suggests that with regard to *public discourses*, schools, media and other forums of public discourse need to be better engaged to inform and educate the public and empower them to participate in governing desertification. In particular, land managers need to be better informed about the extent and symptoms of desertification in their local area.

We have seen throughout this book that the key problem for combating desertification is the existence and use of multiple definitions of 'desertification'. This means that political, scientific and public discourses on desertification need to be brought together in the public sphere, in order to reach a negotiated consensus on what desertification entails in a local area. This also has repercussions for the 'moral codes' of land management which, as Chapters 4-7 have outlined in detail, are currently partly out of sink in our case study areas with the environmental 'realities' on the ground - leading to unsustainable forms of landuse. As a result, ideas of what is justifiable, 'the wrong and right of land management', and underlying the selection of the goals and means of landuse, need to be acknowledged and discussed more openly. As Sullivan (2000) suggests, an important solution will be the *contextualisation of desertification discourses* and revealing the institutionalisation of the 'truths' about desertification, as attempted in this study. Similar approaches are required to bridge the gap between structure and action, and, in particular, to understand the interplay of different interests and structural conditions in the implementation of policies guiding landuse decision-making in general, and land degradation and desertification more specifically.

The policy environment - as the key driver for landuse change in Southern Europe - needs to be substantially altered. To be sure, there are some promising developments that need to be mentioned. For example, the EU 'strategy on soil protection' contains comprehensive requirements for addressing the soil-related dimensions of desertification, although implementation of these requirements needs to be improved and tightened. Further, the *European Spatial Development Perspective* could act as an incentive to develop a more transparent and comprehensive planning process for land management. Nonetheless, our results have emphasised the complexity of the current EU and national policy environment influencing landuse change in Southern Europe, with its many *implicit* and *explicit* effects on desertification processes. In this context, we argue that the 'negative' policy outcomes emanating from this disparate policy arena largely result from the *lack of a unified and coherent EU desertification policy*, as well as from lack of authority of environmental experts and administrators. This has resulted in piecemeal policy approaches in which policy attempts to mitigate desertification have been spread over a wide array of often uncoordinated EU and national policy domains. We argue, therefore, that a unified and holistic *EU-based desertification policy framework* should be put in place that *directly* addresses desertification problems and that brings together the various, currently rather disparate and disconnected, policies that affect desertification processes in often negative ways.

Our evidence also suggests that creating new institutions and developing the processes of policy and implementation, as well as working towards empowering marginal actors through a more aware public discourse, are essential components of all policy change. Participatory approaches in policy design and in achieving the shift towards post-productivist or multifunctional thinking (Wilson, 2001), and the integration of environmental considerations, are also required in order to enhance ownership of new policies, programmes and ideas. Particular attention needs to be paid to the representation of different interest groups in this participation. It is clear from our Italian case study that power in the networks may evolve 'naturally', not least when, due to environmental limitations, the natural resources cease to 'empower' the actors in the same way. However, with the current lack of intervention, the discursive barriers supported by central actors will continue to prevent both the implementation of 'positive' mechanisms for desertification mitigation and the achievement of institutional change.

The results of this research, thus, provide a broad understanding of the policy and socio-economic issues linked to desertification in Southern Europe, and provide tangible suggestions for the development of improved policy implementation processes. By outlining both the structures of actor networks and practices that require changes, and by highlighting the crucial role of the cultural and environmental context in 'unravelling' desertification, results from this book should be of particular importance for policy-makers involved in drafting and implementing NAPs. Indeed, understanding the importance of different stakeholder interpretations of desertification and existing policy agendas could prove particularly crucial for successful implementation of the NAPs at the sub-national level.

At the same time, however, we also acknowledge that much more social science-oriented work on desertification is needed, and we see this book as a basis for further academic investigation into the complex socio-cultural and policy-related processes relevant to land management and desertification - especially with regard to future work involving local stakeholders in desertification-affected areas of Southern Europe. The results discussed here, therefore, hold implications not only for policy formation and implementation, but also for the type of research required to both address the social and political dimensions of desertification and to inform associated policy measures - in other words, there is a continuing need for thickly textured empiricism in social science enquiry of desertification.

References

Abram, S. 1998: Class, countryside and the 'longitudinal study': a response to Hoggart. *Journal of Rural Studies* 14 (3): 369-379.

Abram, S. 1999: Up the anthropologist: power, subversion and progress (another reply to Hoggart). *Journal of Rural Studies* 15 (1): 119-120.

Adger, N., Benjaminsen, A., Brown, K. and H. Svarstad 2001: Advancing a political ecology of global environmental discourses. *Development and Change* 32: 681-715.

Adger, N., Brown, K., Fairbass, J., Jordan, A., Paavola, J., Rosendo, S., and G. Seyfang 2003: Governance for sustainability: towards a 'thick' analysis of environmental decision making. *Environment and Planning A* 35: 1095-1110.

Albiac, J., Tapia, J. and E. Calvo 2002: El uso agrario del agua en las comarcas de Levante y Sureste y el trasvase del Ebro. *Estudios Agrosociales y Pesqueros* 196: 95-131.

Albromeit, H. 1998; How to democratise multi-level, multi-dimensional polity. In: Weale, A. and M. Nentwich (eds): *Political theory and the European Union: legitimacy, constitutional choice and citizenship*. London: Routledge, pp. 112-124.

Albuquerque, J.P. 1961: *Linhas mestras de zonagem climática Portuguesa*. Alcobaça (Portugal): Topografia Alcobacense.

Anania G., Gaudio F. and G. Gaudio 1992: Differenziazioni aziendali, pluriattività domanda, offerta e 'consumo' di politiche nel Mezzogiorno. In: INEA [Istituto Nazionale di Economia Agraria] (eds): *Strategie familiari, pluriattività e politiche agrarie*. Bologna: Il Mulino, pp. 239-314.

Andersen, E., Rutherford, A. and M. Winter 2000: The beef regime. In: Brouwer, F. and P. Lowe (eds): *CAP regimes and the European countryside*. Wallingford: CAB International, pp. 55-70.

Anthopoulou T. and D. Goussios 1994: Transformation rurale et dynamique ovine dans les îles Egéennes (Grèce). In: Centre International des Hautes Etudes Agronomiques Méditerranéennes (ed): *Proceedings of the Second Symposium on livestock farming systems : the study of livestock farming systems in a research and development framework, held in Saragossa (Spain), 11-12 September 1992*. Wageningen: Centre International des Hautes Etudes Agronomiques Méditerranéennes, pp. 79-82.

Arcieri, M., Ferraretto, D., Lamoglie, C. and A. Povellato 2002: Report on the results of in-depth qualitative interviews with local, regional and national policy-makers on the implementation and effects of past policies in the Val d'Agri Target area. In: Juntti, M. and G.A. Wilson (eds): *MEDACTION Project Deliverable 19: National team reports discussing the results of the in-depth qualitative interviews with local, regional and national policy-makers on the implementation and effects of past policies in each of the four target areas*. Maastricht (NL): International Centre for Integrative Studies, pp. 92-127.

Arianoutsou-Faraggitaki, M. 1985: Desertification by overgrazing in Greece: the case of Lesvos island. *Journal of Arid Environments* 9: 237-242.

Ashworth, S. (ed) 2000: *An evaluation of the Common Organisation of the Markets in the sheep and goatmeat sector*. Brussels: Directorate General for Agriculture of the European Commission.

Ashworth, S. and H. Caraveli 2000: The sheepmeat and goatmeat regime. In: Brouwer, F. and P. Lowe (eds): *CAP regimes and the European countryside*. Wallingford: CAB International, pp. 71-86.

Atchia, M. and S. Tropp (eds) 1995: *Environmental management: issues and solutions*. Chichester: Wiley.

Barberá, G.G., López-Bermúdez, F. and A. Romero Díaz 1997: Cambios de uso del suelo y desertificación en el Mediterráneo: el caso del sureste ibérico. In: García Ruiz, J.M. and P. López García (eds): *Acción humana y desertificación en ambientes mediterráneos*. Zaragoza (Spain): Instituto Pirenaico de Ecología CSIC, pp. 9-39.

Basso F., Bove E., Del Prete M. and M. Pisante 1998: *The Agri Basin, Basilicata, Italy*. In: Mairota, P., Thornes, J.B. and N. Geeson 1998: *Atlas of Mediterranean environments in Europe: the desertification context*. Chichester: Wiley, pp. 144-151.

Baxter, J. and J. Eyles 1997: Evaluating qualitative research in social geography: establishing 'rigour' in interview analysis. *Transactions of the Institute of British Geographers* 37 (4): 505-525.

Beck, U. 1992: *Risk society*. London: Sage.

Beopoulos N., 1997: L'intensification de l'agriculture grecque et les problèmes de l'environnement. *Options Méditerranéennes B* 12: 217-224.

Beopoulos, N. 2002: Mediterranean agriculture in the light and shadow of the CAP. In: Kasimis, C. and G. Stathakis (eds): *The reform of the CAP and rural development in Southern Europe*. Aldershot: Ashgate, pp. 35-48.

Beopoulos, N., Sergianni, D. and G. Vlahos 2002: Report on the results of in-depth qualitative interviews with local, regional and national policy-makers on the implementation and effects of past policies in the Lesvos Target Area. In: Juntti, M. and G.A. Wilson (eds): *MEDACTION Project Deliverable 19: National team reports discussing the results of the in-depth qualitative interviews with local, regional and national policy-makers on the implementation and effects of past policies in each of the four target areas*. Maastricht (NL): International Centre for Integrative Studies, pp. 49-71.

Beopoulos, N., Vlahos, G. and D. Sergianni 2003: Greece: report on the farm questionnaire survey. In: Wilson, G.A. and M. Juntti (eds): *MEDACTION Project Deliverable 20: A report by each national team discussing the results of the farm questionnaire survey*. Maastricht (NL): International Centre for Integrative Studies, pp. 45-80.

Blaikie, P. and H. Brookfield 1987: *Land degradation and society*. London: Methuen.

Boenzi F. and R. Giura Longo 1994: *La Basilicata: i tempi, gli uomini, l'ambiente*. Bari (Italy): Edipuglia.

Börzel, T.A. 2000: Why there is no southern problem: on environmental leaders and laggards in the European Union. *Journal of European Public Policy* 7 (1): 141-162.

Botkin, D.B. and E.A. Keller 1998: *Environmental science: Earth as a living planet* (2nd edition). Chichester: Wiley.

Boutonnet, J.P. 1993: Les revenus des éleveurs ovins. In: Société Française d'Economie Rurale (ed): *Les revenues agricoles*. Proceedings of a conference held at Montpellier (France), 13-14 May 1993. Paris: Société Française d'Economie Rurale, pp. 1-9.

Bowyer-Bower, T. 2003: Desertification in East and Southern Africa. In: Potts D. and T. Bowyer-Bower (eds): *Eastern and Southern Africa: development challenges in a volatile region*. Harlow: Pearsons, pp. 229-254.

Brandt, C.J. and J.B. Thornes 1996: *Mediterranean desertification and land use*. Chichester: Wiley.

Briassoulis H., in collaboration with M. Juntti and G. Wilson 2003: *Mediterranean desertification: framing the policy context*. Brussels: European Commission (Report EUR 20731).

Briassoulis, H. (in press): The institutional complexity of environmental policy and planning problems: the example of Mediterranean desertification. *Journal of Environmental Planning and Management*.

Brouwer, F. and P. Lowe (eds) 1998: *CAP and the rural environment in transition: a panorama of national perspectives*. Wageningen (NL): Wageningen Pers.

Bruckmeier, K. and T. Patricio 2002: The agri-environmental policy of the European Union: new chances for development in the south European countryside? In: Kasimis, C. and G. Stathakis (eds): *The reform of the CAP and rural development in Southern Europe*. Aldershot: Ashgate, pp. 49-70.

Buller, H. 2000: Regulation 2078: patterns of implementation. In: Buller, H., Wilson, G.A. and A. Höll (eds): *Agri-environmental policy in the European Union*. Aldershot: Ashgate, pp. 219-253.

Buller, H. 2001: Is this the European model? In: Buller H. and K. Hoggart (eds): *Agricultural transformation, food and environment: perspectives on European rural policy and planning (Vol 1)*. Aldershot: Ashgate, pp. 1-8.

Buller, H., Wilson, G.A. and A. Höll (eds) 2000: *Agri-environmental policy in the European Union*. Aldershot: Ashgate.

Buttel, F. 2001: Some reflections on late twentieth century agrarian political economy. *Sociologia Ruralis* 41 (2): 165-181.

Cafiero, C., Cioffi, A., Cupo, P., Pomarici, E. and F. Sallusti 1998: Small farms in southern Italy. In: Monke E., Avillez F. and S. Pearson (eds): *Small farm agriculture in Southern Europe: CAP reform and structural change*. Aldershot: Ashgate, pp. 65-122.

Calatrava, J. 2004: Cross-compliance for soil erosion control in semi-arid regions. Unpublished paper presented at the Seminar 'Evaluation of cross-compliance' as part of the EU Concerted Action Project *Developing cross compliance in the EU: background, lessons and opportunities* (QLK5 - CT- 2002-02640), 19-20 April 2004, Granada (Spain).

Callon, M. 1986: Some elements of a sociology of translation: domestication of the scallops and the fishermen of St. Brieuc Bay. In: Law, J. (ed): *Power, action and belief: a new sociology of knowledge?* London: Routledge and Kegan Paul, pp. 196-223.

Callon, M. 1999: Actor Network Theory: the market test. In: Law, J. and J. Hassard (eds): *The actor-network theory and after*. Oxford: Blackwell, pp. 181-195.

Cannata, G. and M. Reho 1981: Le politiche settoriali di sviluppo dell'agricoltura e la pianificazione territoriale. *Rivista di Economia Agraria* 36 (2): 345-68.

Casadei, E. 1991: Attività produttiva agraria e tutela del paesaggio: profili giuridici. *Agricoltura e Paesaggio (Quaderno dell' Accademia dei Georgofili)* 4: 35-56.

Castree, N. and B. Braun 1998: The construction of nature and the nature of construction: analytical and political tools for building survivable futures. In: Braun, B. and N. Castree (eds): *Remaking reality: nature at the millennium*. London: Routledge, pp. 3-42.

CEC [Commission of the European Communities] 1992: *CORINE: soil erosion risk and land resources in the southern regions of the European Community*. Luxembourg: European Commission (Report EUR 13233).

CEC [Commission of the European Communities] 1997: *International Conference on Mediterranean Desertification: research results and policy implications*. Luxembourg: European Commission (Report EUR 17782 EN).

CEC [Commission of the European Communities] 2000: *Addressing desertification and land degradation: the activities of the European Community in the context of the United Nations Convention to Combat Desertification*. Brussels: European Commission.

CEC [Commission of the European Communities] 2001: Communication from the Commission to the Council of the European Parliament, the Economic and Social Committee and the Committee of Regions on the sixth environmental action programme of the European Community 'Environment 2010: Our Future, Our Choice'. Available online at http://europa.eu.int/eur-lex/en/com/pdf/2001/en_501PC 0031.pdf

CEC [Commission of the European Communities] 2002a: *Towards a thematic strategy for soil protection*. Brussels: CEC (Report COM[2002] 179 final).

CEC [Commission of the European Communities] 2002b: *Life III: the financial instrument for the environment*. Luxembourg: Office for the Official Publications of the European Communities.

CEH [Consejería de Economía y Hacienda] (several years): *Anuario estadístico de la Región de Murcia*. Murcia (Spain): CEH.

Cembalo, A. 1980: Analisi della convenienza degli allevamenti zootecnici nelle zone interne del Mezzogiorno continentale. In: De Benedictis M. (ed): *L'agricoltura nello sviluppo del Mezzogiorno*. Bologna: Il Mulino, pp. 157-185.

Cerdá, A. 1997: Soil erosion after land abandonment in a semiarid environment of southeastern Spain. *Arid Soil Research and Rehabilitation* 11: 163-176.

CESRM [Consejo Económico y Social de la Región de Murcia] 1997: *Informe sobre el sector hortofrutícola ante la reforma de la OCM de frutas y hortalizas frescas*. Murcia (Spain): CESRM (Informe 1/1997).

Chaparro, J. and M.A. Esteve 1995: Evolución geomorfológica de laderas repobladas mediante aterrazamientos en ambientes semiáridos (Murcia, SE de España). *Cuaternario y Geomorfología* 9: 34-49.

CHS [Confederación Hidrográfica del Segura] 2001: *Plan hidrológico de la Cuenca del Segura*. Murcia (Spain): CHS-Ministerio de Medio Ambiente.

Cianferoni, R., Fattori, M. and M. Zoppi Spini 1976: L'agricoltura nei piani regolatori comunali. *Rivista di Economia Agraria* 31 (4): 689-703.

Cioffi, A. 1997: Le prospettive delle piccole aziende nelle zone rurali del Mezzogiorno interno: l'Alta Val d'Agri. In: Cioffi, A. and A. Sorrentino (eds): *La piccola azienda e la nuova politica agricola dell'Unione Europea*. Milano: Franco Angeli, pp. 169-199.

Clark, C. 1974: *Population, growth and land use*. London: McMillan.

Clark, N. and F. Perez-Trejo 1995: The interaction between research and policy in the management of desertification. In: Fantechi, R., Peter, D., Balabanis, P. and J.L. Rubio (eds): *Desertification in a European context: physical and socio-economic aspects*. Proceedings of the European School of Climatology and Natural Hazards Course in Alicante, Spain, 6-13 October 1993. Luxembourg: Office for the Official Publications of the European Communities, pp. 341-355.

Clarke, M.L. and H.M. Rendell 2000: The impact of the farming practice of remodelling hillslope topography on badland morphology and soil erosion processes. *Catena* 40: 229-250.

CoE [Council of Europe] 1972: *European Soil Charter - B(72)63*. Strasbourg: Council of Europe.

CoE [Council of Europe] 2002: *Draft revised Soil Charter - CO-DBP (2002)2*. Strasbourg: Council of Europe.

Colomer, J.C., Morell, C., Sanches, J., Boluda, R. and A. Artiago 1995: A study of soil degradation within the EFEDA project. In: Fantechi, R., Peter, D., Balabanis, P. and J.L. Rubio (eds): *Desertification in a European context: physical and socio-economic aspects*. Proceedings of the European School of Climatology and Natural Hazards Course in Alicante, Spain, 6-13 October 1993. Luxembourg: Office for the Official Publications of the European Communities, pp. 447-456.

Comissão Nacional do Ambiente 1982: *Carta ecológica de Portugal: atlas do ambiente*. Lisbon: Instituto do Ambiente.

Conforti, P. 2002: *Un'analisi delle ipotesi di riduzione del sostegno al grano duro nell'Unione Europea e dei suoi effetti in Italia*. Rome: INEA (Working Paper no. 14).

COP4 [Conference of the Parties] 2000: *Implementation of the UNCCD, part one: synthesis of the information contained in national report from Northern Mediterranean and other affected country parties*. Bonn: ICCD/COP4.

Cornet, A. 2002: Desertification and its relationship to the environment and development: a problem that affects us all. In: Barbault, R., Cornet, A., Jouzel, J., Mégie, G., Sachs, I. and J. Weber (eds): *Johannesburg World Summit on Sustainable Development 2002: what is at stake? The contribution of scientists to the debate*. Paris: Ministère des Affaires étrangères, pp. 91-125.

Cozza, F. 2000: *Alcune considerazioni di carattere generale, con particolare riferimento alla gestione dei boschi ed al quadro normativo: suggerimenti e proposte*. www.desertification.it

CRIC1 [Committee for the Review of the Implementation of the Convention] 2003: *Report of the Committee from its first session held in Rome from 11 to 22 of November 2002*. Bonn: ICCD/CRIC/UN.

Crosby, B. 1996: Policy implementation: the organizational challenge. *World Development* 24 (9): 1403-1415.

Cummings, C., Gómez, A., Oñate, J.J., Peco, B. and J.M. Sumpsi 2001a: Report on policy work already conducted in the Guadalentín area under previous projects. In: Jenkins, J. and G.A. Wilson (eds): *MEDACTION Project Deliverable 15: National team reports on policy work already conducted in research areas under previous projects*. Maastricht (NL): International Centre for Integrative Studies, pp. 15-21.

Cummings, C., Oñate, J.J., Gómez, A., Peco, B. and J.M. Sumpsi 2001b: Report on past and present policies which should form the focus of investigation in the Guadalentín target area. In: Jenkins, J. and G.A. Wilson (eds): *MEDACTION Project Deliverable 18: National team reports on past and present policies which should form the focus of investigation*. Maastricht (NL): International Centre for Integrative Studies, pp. 13-23.

Dax, T. and P. Hellegers 2000: Policies for Less Favoured Areas. In: Brouwer, F. and P. Lowe (eds): *CAP regimes and the European countryside*. Wallingford: CAB International, pp. 179-198.

De Benedictis, M. 2002: L'agricoltura del Mezzogiorno: 'la polpa e l'osso' cinquant'anni dopo. *La Questione Agraria* 2: 199-236.

De Benedictis, M. 2003: La terra dell'osso. *La Questione Agraria* 4: 119-25.

De Stefano, F. 2003: Il Mezzogiorno agricolo e 'l'osso': qualcosa si muove? *La Questione Agraria* 1: 31-51.

Demeritt, D. 1996: Social theory and the reconstruction of science and geography. *Transactions of the Institute of British Geographers* 21 (3): 484-503.

Demeritt, D. 2001: Being constructive about nature. In: Castree N. and B. Braun (eds): *Social nature: theory, practice and politics*. Oxford: Blackwell, pp. 22-40.

Diamandouros, P.N. and R. Gunther 2001: *Parties, politics and democracy in the new Southern Europe*. London: John Hopkins University Press, pp. 1-15.

Diniz, M.A. 1997: *A conservação da natureza na política agrícola portuguesa*. Lisbon: Universidade Técnica de Lisboa.

Dono, G. 1996: Una politica ambientale per l'agricoltura collinare dell'Italia Centrale. *La Questione Agraria* 64: 57-81.

Dregne, H., Kassas, M. and B. Rozanov 1991: A new assessment of the world status of desertification. *Desertification Control Bulletin* 20: 6-18.

Eboli, M.G. 1992: Contesti socio-economici, pluriattività e tipologie di aziende-famiglia: una analisi comparata. In: INEA [Istituto Nazionale di Economia Agraria] (eds): *Stategie familiari, pluriattività e politiche agrarie*. Bologna: Il Mulino, pp. 57-100.

Eden, P. 2002: Personal communication, October 2002, Instituto para o Desenvolvimento Rural e Gestão, Lisbon, Portugal.

Eder, K. 1996: *The social construction of nature*. London: Sage.

EEA [European Environment Agency] 2000: *Down to Earth: soil degradation and sustainable development in Europe*. Copenhagen: EEA.

El País 1999: El Segura, una cloaca. *El País* (Valencia), 3rd June 1999: 29.

Eriksen, S. 2001: Linkages between climate change and desertification in East Africa. *Arid Lands Newsletter* 49. Available online at: http://ag.arizona.edu/OALS/ALN/aln49

Escartín, C.M. and J.M. Santafé 2001: Application of the cost recovery principle in Spain: policies and impacts. In: European Commission (eds): *Pricing water: economics, environment and society*. Brussels: European Commission, pp. 131-140.

Escobar, A. 1998: Whose knowledge, whose nature? Biodiversity, conservation and the political ecology of social movements. *Journal of Political Ecology* 5: 53-82.

Esteve, M.A., Ferrer, D., Ramirez Díaz, L., Calvo, J.F., Suárez Alonso, M.L. and M.R. Vidal Abarca 1990: Restauración de la vegetación en ecosistemas áridos y semiáridos: alguinas reflexiones ecológicas. *Ecología* 1: 497-510.

European Commission 2000: *Addressing desertification and land degradation: the activities of the European Community in the context of the United Nations Convention to Combat Desertification*. Brussels: European Commission.

Fabiani, G. 1986: *L'agricoltura italiana tra sviluppo e crisi (1945-1985)*. Bologna: Il Mulino.

Fairhead, J. and M. Leach 1996: *Reversing landscape history: power, policy and socialised ecology in West Africa's forest-savanna mosaic*. Chichester: Wiley.

Falconer, K. and N. Ward 2000: *Modulation and the implementation of the Common Agricultural policy Reform in the UK*. Newcastle: Centre for Rural Economy (CRE Working Paper 44).

Fanfani, R. 1990: *Lo sviluppo della politica agricola comunitaria*. Rome: La Nuova Italia Scientifica.

Fantechi, R. and N.S. Margaris (eds) 1986: *Desertification in Europe*. Dordrecht (NL): Reidel.

References

Fantechi, R., Peter, D., Balabanis, P. and J.L. Rubio (eds) 1995: *Desertification in a European context: physical and socio-economic aspects*. Proceedings of the European School of Climatology and Natural Hazards Course in Alicante, Spain, 6-13 October 1993. Luxembourg: Office for the Official Publications of the European Communities.

FAO [Food and Agriculture Organisation] 1977: *World map of desertification*. Nairobi: UNCOD.

Feio, M. 1993: O colapso dos cereais. *Gazeta das Aldeias* 3011 (Universidade de Evora): 18-23.

Ferraretto D., Povellato A. and M. Arcieri 2003: Italy: report of farm questionnaire survey. In: Wilson, G.A. and M. Juntti (eds): *MEDACTION Project Deliverable 20: A report by each national team discussing the results of the farm questionnaire survey*. Maastricht (NL): International Centre for Integrative Studies, pp. 81-114.

Ferro, O. 1988: *Istituzioni di politica agraria*. Bologna: Edagricole.

Ficco, P., Rifici, R. and M. Santoloci 1999: *La nuova tutela delle acque: gli obblighi, gli obiettivi e gli strumenti previsti dal Dlgs 152/1999*. Milano: Edizioni ambiente.

Fino, M.L. 1993: Medidas agro-ambientais: sua aplicação em Portugal. *Revista da Agricultura/ CAP* 53: 33-38.

Formica, C. 1975: *Lo spazio rurale nel Mezzogiorno: esodo, desertificazione e riorganizzazione*. Napoli: Edizioni Scientifiche Italiane.

Foucault, M. 1991: Politics and the study of discourse. In: Burchell, G., Colon, G. and P. Miller (eds): *The Foucault effect: studies in governmentality*. London: Harvester Wheatsheaf, pp. 53-72.

Frouws, J. and A.P. Mol 1997: Ecological modernization theory and agricultural reform. In: De Haan, H. and N. Long (eds): *Images and realities of rural life*. Wageningen (NL): Wageningen Pers, pp. 269-286.

Garrido, F. and E. Moyano 1996: Spain. In: Whitby, M. (ed): *The European environment and CAP reform: policies and prospects for conservation*. Wallingford: CAB International, pp. 86-104.

Geeson, N.A., Brandt, C.J. and J.B. Thornes (eds) 2002: *Mediterranean desertification: a mosaic of processes and responses*. Chichester: Wiley.

Giourga, C. 1991: The change of the traditional land management model in the Archipelago of the Aegean: repercussions for island ecosystems (original in Greek). Unpublished PhD thesis, University of the Aegean, Lesvos.

Gisotti, G. and M. Benedini 2000: *Il dissesto idrogeologico: prevenzione, prevenzione e mitigazione del rischio*. Roma: Carocci.

Glenn, E., Stafford Smith, M. and V. Squires 1998: On our failure to control desertification: implications for global change issues and a research agenda for the future. *Environmental Science and Policy* 1: 71-78.

Goodman, D. 2001: Ontology matters: the relational materiality of nature and agro-food studies. *Sociologia Ruralis* 41 (2): 182-200.

Gorgoni, M. 1980: Agricoltura contadina e migrazioni temporanee nel Mezzogiorno interno. In: De Benedictis, M. (ed): *L'agricoltura nello sviluppo del Mezzogiorno*. Bologna: Il Mulino, pp. 67-86.

Graziani, A. 1998: *Lo sviluppo dell'economia italiana: dalla ricostruzione alla moneta unica*. Torino: Bollati Boringhieri.

Grillenzoni, M. and A. Ragazzoni 1995: Le aree protette per la conservazione attiva dell'ambiente: inquadramento metodologico. *Studi di Economia e Diritto* 40 (1): 59-83.

Guariglia, A. 1994: *Politica agraria e legislazione (1944-1992)*. Roma: Elea Press.

Hagedorn, K. (ed) 2002: *Environmental co-operation and institutional change: theories and policies for European agriculture*. Cheltenham: Edward Elgar.

Hannigan, J.A. 1995: *Environmental sociology: a social constructionist perspective*. London: Routledge

Harrison, J. 1993: *The Spanish economy: from civil war to the European Community*. Basingstoke: Macmillan.

Hart, K. and G.A. Wilson 1998: UK implementation of Agri-environment Regulation 2078/92/EEC: enthusiastic supporter or reluctant participant? *Landscape Research* 23: 255-272.

Helldén, U. 1991: Desertification: time for an assessment. *Ambio* 20: 372-383.

Hertin, J. and F. Berkhout 2003: Analysing institutional strategies for environmental policy integration: the case of EU enterprise policy. *Journal of Environmental Policy and Planning* 5 (1): 39-56.

Hildebrand, P.M. 2002: The European Community's Environmental policy 1957 to 1992: from incidental measures to an international regime? In: Jordan, A. (ed): *Environmental policy in the European Union: actors, institutions and processes.* London: Earthscan, pp. 13-36.

Hill, M. and P. Hupe 2002: *Implementing public policy.* London: Sage.

Hoggart, K. 1990: Let's do away with rural. *Journal of Rural Studies* 6: 245-257.

Hoggart, K. 1998: Rural cannot be middle class because class does not exist? *Journal of Rural Studies* 13 (3): 253-273.

Hudson, R. 1999: Putting policy into practice: policy implementation problems, with special reference to the European Mediterranean. In: Balabanis, P., Peter, D., Ghazi, A. and M. Tsogas (eds): *Mediterranean desertification: research results and policy implications* (Vol 1). Luxembourg: Office for Official Publications of the European Communities, pp. 243-253.

Imeson, A.C. and L.H. Cammeraat 2002: Environmentally Sensitive Areas in the MEDALUS target area study sites. In: Geeson, N.A., Brandt, C.J. and J.B. Thornes (eds) 2002: *Mediterranean desertification: a mosaic of processes and responses.* Chichester: Wiley, pp. 177-186.

INAG [Instituto da Agua] 2000: Instituto da Agua web pages. http://www.inag.pt/

INE [Instituto Nacional de Estadística] 2000ª: *Contabilidad regional de España.* Madrid: INE.

INE [Instituto Nacional de Estadística] 2000b: *Censo agrario 1999.* Madrid: INE.

INE [Instituto Nacional de Estadística] 2002: *Censo de población 2001.* Madrid: INE.

INE [Institutio Nacional de Estatistica] 1999: Institutio Nacional de Estatistica Portugal web pages. http://www.ine.pt/index.htm

INEA [Istituto Nazionale di Economia Agraria] 1999: *Le misura agroambientali in Italia: analisi e valutazione del reg. CEE 2078/92 nel quadriennio 1994-97 (Rapporto nazionale).* Rome: Istituto Nazionale di Economia Agraria.

Izcara Palacios, S.P. 1998: Farmers and the implementation of the EU Nitrates Directive in Spain. *Sociologia Ruralis* 38: 146-162.

Jones, A. and J. Clark 2001: *The modalities of European Union governance: New Institutionalist explanations of agri-environmental policy.* Oxford: Oxford University Press.

Jordan, A. 2002: Introduction: European Union environmental policy - actors, institutions and policy processes. In: Jordan, A. (ed): *Environmental policy in the European Union: actors, institutions and processes.* London: Earthscan, pp. 1-10.

Juntti, M. and C. Potter 2002: Interpreting and re-interpreting agri-environmental policy: communication, trust and knowledge in the implementation process. *Sociologia Ruralis* 42 (3): 215-232.

Juntti, M. and G.A. Wilson (in press): Conceptualising desertification in Southern Europe: stakeholder interpretations and multiple policy agendas. *European Environment* 15.

Kinlund, P. 1996: *Does land degradation matter? Perspectives on environmental change in northeastern Botswana.* Stockholm: Almqvist and Wicksell International.

Kontis, I. 1978: *Lesvos and the Asia Minor region: ancient greek cities* (in Greek). Athens: Athens Centre of Ekistics.

Kosmas, C., Danalatos, N.G., Cammeraat, L.H., Chabart, M., Diamantopoulos, J., Farand, R., Gutierrez, L., Jacob, A., Marques, H., Martinez-Fernandez, J., Mizara, A., Moustakas, N., Nicolau, J.M., Oliveros, C., Pinna, G., Puddu, R., Puigdefabregas, J., Roxo, M.J., Simao, A., Stamou, G., Tomasi, N., Usai, D. and A. Vacca 1997: The effect of land use on runoff and soil erosion rates under Mediterranean conditions. *Catena* 29: 45-59.

Kosmas, C., Danalatos, N. and S. Gerontidis 1999a: Land characteristics and management practices affecting desertification in Lesvos. In: Balabanis, P., Peter, D., Ghazi, A. and M. Tsogas (eds): *Mediterranean desertification: research results and policy implications* (Vol 1). Luxembourg: Office for Official Publications of the European Communities, pp. 77-86.

Kosmas, C., Gerontidis, S., Detsis, V., Zafiriou, T. and M. Marathianou 1999b: The island of Lesvos. In: Kosmas, C., Kirkby, M. and N. Geeson (eds): *Manual on key indicators of desertification and mapping environmentally sensitive areas to desertification*. Luxembourg: Office for Official Publications of the European Communities, pp. 66-73.

Kosmas, C., Kirkby, M. and N. Geeson (eds) 1999c: *Manual on key indicators of desertification and mapping environmentally sensitive areas to desertification*. Luxembourg: Office for Official Publications of the European Communities.

Kosmas, C., Yassoglou, N., Danalatos, N., Karavitis, C., Gerontidis, S., and A. Mizara 1996: Lesvos: land degradation and desertification. In: Brandt, C.J. and J.B. Thornes 1996 (eds): *Mediterranean desertification and land use: MEDALUS III first annual report*. London: King's College London, pp. 204-215.

Kousis, M. 2000: A political economy approach to the 'Mediterranean Syndrome'. In: Kousis, M. and J. Lekakis (eds): *Coming to terms with the 'Mediterranean Syndrome': the implementation of European environmental policies in southern states*. Florence (Italy): European University Institute.

La Opinión 2001: La Comunidad dice que suprime zonas de espacios naturales por no tener valor. *La Opinión* (Murcia), 24th March 2001: 8.

La Spina, A. and G. Sciortino 1993: Common agenda, southern rules: European integration and environmental change in the Mediterranean states. In: Liefferink, J.D., Lowe. P. and A.P. Mol (eds): *European integration and environmental policy*. London: Belhaven, pp. 217-236.

Lal, R. 2001: Soil degradation by erosion. *Land Degradation and Development* 12 (6): 119-139.

Latour, B. 1986: The powers of association. In: Law, J. (ed): *Power, action and belief: a new sociology of knowledge*. London: Routledge, pp. 264-280.

Latour, B. 1993: *We have never been modern*. Cambridge (Mass.): Harvard University Press.

Law, J. 1992: Notes on the theory of the actor network: ordering, strategy and heterogeneity. http://www.comp.lancs.ac.uk/sociology/soc054jl.html

Law, J. 1999: After ANT: complexity, naming and topology. In: Law, J. and J. Hassard (eds): *The actor-network theory and after*. Oxford: Blackwell, pp. 1-14.

Law, J. and J. Hassard (eds) 1998: *The actor-network theory and after*. Oxford: Blackwell.

Leach, M. and R. Mearns 1996: Environmental change and policy: challenging received wisdom in Africa. In: Leach, M. and R. Mearns (eds): *The lie of the land: challenging received wisdom on the African environment*. Oxford: Heinemann, pp. 1-33.

Lemon, M. (ed) 1999: *Exploring environmental change using an integrative method*. Reading: Gordon and Breach.

Lemon, M., Seaton, R. and J. Park 1994: Social enquiry and the measurement of natural phenomena: the degradation of irrigation water in the Argolid Plain, Greece. *International Journal of Sustainable Development and World Ecology* 2 (3): 206-220.

Levi, C. 1945: *Cristo si è fermato a Eboli*. Torino: Einaudi.

Liefferink, J.D., Lowe, P. and A.P. Mol 1993: The environment and the European Community: the analysis of political integration. In: Liefferink, J.D., Lowe. P. and A.P. Mol (eds) *European integration and environmental policy*. London: Belhaven, pp. 1-14.

Lipsky, M. 1980: *Street level bureaucracy: dilemmas of the individual in public services*. New York: Russel Sage Foundation.

Long, N. 1992: From paradigm lost to paradigm gained? The case of an actor oriented sociology of development. In: Long, N. and A. Long (eds): *Battelfields of knowledge: the interlocking of theory and practice in social research and development*. London: Routledge, pp. 16-43.

Long, N. 1997: Agency and constraint, perceptions and practice. In: De Haan, H. and N. Long (eds): *Images and realities of rural life*. Wageningen (NL): Wageningen Pers, pp.1-22.

Long, N. and A. Long 1992: *Battelfields of knowledge: the interlocking of theory and practice in social research and development*. London: Routledge.

López-Bermúdez, F., Barberá, G.G., Alonso Sarriá, F. and F. Belmonte 2002: Natural resources in the Guadalentín basin (SE Spain): water as a key factor. In: Geeson, N.A., Brandt, C.J. and J.B. Thornes (eds): *Mediterranean desertification: a mosaic of processes and responses*. Chichester: Wiley, pp. 233-246.

López-Bermúdez, F., Romero-Díaz, A., Cabezas, F., Rojo-Serrano, L., Martínez-Fernández, J., Böer, M. and G. Del Barrio 1998: The Guadalentín basin, Murcia, Spain. In: Mairota, P., Thornes, J.B. and N. Geeson (eds): *Atlas of Mediterranean environments in Europe: the desertification context*. Chichester: Wiley, pp. 130-142.

Loureiro, N.S., Campos, N. and J.M. Pereira 1994: *Medidas para a conservaçao e recuperaçao de solos no Algarve*. Faro (Portugal): Universidade de Aveiro.

Low, N. 2002: Ecosocialisation and environmental planning: a Polanyian approach. *Environment and Planning A* 34: 43-60.

Lowe, P. and F. Brouwer 2000: Agenda 2000: a wasted opportunity? In: Brouwer, F. and P. Lowe (eds): *CAP regimes and the European countryside*. Wallingford: CAB International, pp. 321-334.

Lowe, P., Buller, H. and N. Ward 2002: Setting the next agenda? British and French approaches to the second pillar of the Common Agricultural Policy. *Journal of Rural Studies* 18: 1-18.

Lowe, P., Clark, J., Seymour, S. and N. Ward 1997: *Moralizing the environment: countryside change, farming and pollution*. London: UCL Press.

Lowndes, V. and D. Wilson 2003: Balancing revisability and robustness: a New Institutionalist perspective on local government modernization. *Public Administration* 81 (2): 275-298.

Mairota, P., Thornes, J.B. and N. Geeson 1998: *Atlas of Mediterranean environments in Europe: the desertification context*. Chichester: Wiley.

Mannion, A.M. and S.R. Bowlby (eds) 1992: *Environmental issues in the 1990s*. Chichester: Wiley.

Mantino, F. and E. Turri 1992: Strategie familiari, comportamenti aziendali e politiche agrarie nel Lazio meridionale. In: INEA [Istituto Nazionale di Economia Agraria] (eds): *Strategie familiari, pluriattività e politiche agrarie*. Bologna: Il Mulino, pp. 199-238.

MAPA [Ministerio de Agricultura, Pesca y Alimentación] (several years): *Anuario de estadística agraria*. Madrid: MAPA.

MAPA [Ministerio de Agricultura, Pesca y Alimentación] 2002: *Plan nacional de regadíos: horizonte 2008*. Madrid: MAPA.

Marathianou, M., Kosmas, C., Gerontidis, S. and V. Detsis 2000: Land-use evolution and degradation in Lesvos (Greece): a historical approach. *Land Degradation and Development* 11: 63-73.

Margaris, N.S. 1987: Desertification in the Aegean Islands. *Ekistics* 323/324: 132-136.

Margaris, N.S., Koutsidou, E. and C. Giourga 1996: Changes in traditional Mediterranean land-use systems. In: Brandt, C.J. and J.B. Thornes (eds): *Mediterranean desertification and land use*. Chichester: Wiley, pp. 29-42.

Marini, M. 1995: I fattori che influenzano le scelte aziendali secondo un approccio struttura-agente. In: De Benedictis, M. (ed): *Agricoltura familiare in transizione*. Rome: INEA, pp. 211-254.

Marsden, T. 2000: Food matters and the matter of food: towards a new food governance? *Sociologia Ruralis* 40 (1): 20-29.

Marsden, T. 2004: The quest for ecological modernisation: re-spacing rural development and agri-food studies. *Sociologia Ruralis* 44 (2): 129-146.

Marsden, T., Murdoch, J., Lowe, P., Munton, R. and A. Flynn 1993: *Constructing the countryside*. London: UCL Press.

Martínez, J. and M.A. Esteve 2000: El regadío en la Cuenca del Segura y sus efectos ambientales y sociales. In: Martínez, J. (ed): *Gestión alternativa del agua en la Cuenca del Segura*. Murcia (Spain): Ecologistas en Acción-Región Murciana, pp. 53-70.

Massarutto, A. 1999: *Agriculture, water resources and water policies in Italy*. Milano: Fondazione Eni Enrico Mattei (Working paper no. 33.99).

Mendes, F.L. 1992: A floresta e as pastagens. *Floresta e Ambiente* 16.

References

Mensching, H.G. 1986: Desertification in Europe? A critical comment with examples from the European Mediterranean. In: Fantechi, R. and N.S. Margaris (eds): *Desertification in Europe*. Dordrecht (NL): Reidel, pp. 101-117.

Merlo, M. 1991: The effects of late economic development on land use. *Journal of Rural Studies* 7 (4): 445-457.

Middleton, N. 1995: *The global casino*. London: Edward Arnold.

MIMAM [Ministerio de Medio Ambiente] 2001a: *Plan hidrológico nacional*. Madrid: MIMAM.

MIMAM [Ministerio de Medio Ambiente] 2001b: *Programa de acción nacional contra la desertificación: borrador de trabajo*. Madrid: MIMAM.

MINHAC [Ministerio de Hacienda] 2000: *Plan de desarrollo regional 2000-2006*. Madrid: MINHAC.

Montresor, E. 1994: Il ruolo dell'agricoltura nelle aree protette. *La Questione Agraria* 55: 23-42.

Morris, C. and C. Potter 1995: Recruiting the new conservationists: farmers' adoption of agri-environmental schemes in the UK. *Journal of Rural Studies* 11 (1): 55-63.

Moschini, R. 2000: La legge sulle aree protette 10 anni dopo. *Parchi* 31: 3-9.

Mourao, J.M. 1998: Governmental action to combat desertification in Portugal. In: Burke, S. and J.B Thornes (eds): *Actions taken by national governmental and non-governmental organisations to combat desertification in the Mediterranean*. Brussels: European Commission (Report EUR 18490 EN), pp. 177-205.

Moxey, A., Whitby, M. and P. Lowe 1998: *Environmental indicators for a reformed CAP: monitoring and evaluating policies in agriculture*. Newcastle: Centre for Rural Economy.

Murdoch, J. 1997: The shifting territory of government: some insights from the Rural White Paper. *Area* 29 (2): 109-118.

Murdoch, J. 2000: Networks: a new paradigm of rural development? *Journal of Rural Studies* 16 (4): 407-419.

National Forests Inventory 1995: *Inventario florestal nacional*. Lisbon: Direcçao geral das florestas.

Newby, H. 1980: Rural sociology: a trend report. *Current Sociology* 28 (1): 3-141.

NSSG [National Statistical Service of Greece] 1961a: *National population census* (in Greek). Athens: National Statistical Service of Greece.

NSSG [National Statistical Service of Greece] 1961b: *Agricultural census* (in Greek). Athens: National Statistical Service of Greece.

NSSG [National Statistical Service of Greece] 1971a: *National population census* (in Greek). Athens: National Statistical Service of Greece.

NSSG [National Statistical Service of Greece] 1971b: *Agricultural census* (in Greek). Athens: National Statistical Service of Greece.

NSSG [National Statistical Service of Greece] 1981a: *National population census* (in Greek). Athens: National Statistical Service of Greece.

NSSG [National Statistical Service of Greece] 1981b: *Agricultural census* (in Greek). Athens: National Statistical Service of Greece.

NSSG [National Statistical Service of Greece] 1991a: *National population census* (in Greek). Athens: National Statistical Service of Greece.

NSSG [National Statistical Service of Greece] 1991b: *Agricultural census* (in Greek). Athens: National Statistical Service of Greece.

NSSG [National Statistical Service of Greece] 1991c: *Landuse categories 1991* (in Greek). Athens: National Statistical Service of Greece.

NSSG [National Statistical Service of Greece] 2001: *National population census* (in Greek). Athens: National Statistical Service of Greece.

Nugent, N. 2003: *The government and politics of the European Union*. Basingstoke: Macmillan.

Obando, J.A. 2002: The impact of land abandonment on regeneration of semi natural vegetation: a case study from the Guadalentín. In: Geeson, N.A., Brandt, C.J. and J.B. Thornes (eds): *Mediterranean desertification: a mosaic of processes and responses*. Chichester: Wiley, pp. 269-276.

OECD [Organisation for Economic Cooperation and Development] 2002: *Aid targeting the objectives of the Rio Conventions 1998-2000*. Paris: OECD.

Olsson, L. 1985: *An integrated study of desertification: applications of remote sensing, GIS and spatial models in semi-arid Sudan*. Malmö (Sweden): University of Lund Press.

Oñate, J.J., Cummings, C., Gómez, A., Peco, B. and J.M. Sumpsi 2001: Report on published and unpublished references on policy implementation and desertification at the local, regional and national level in Spain. In: Jenkins, J. and G.A. Wilson (eds): *MEDACTION Project Deliverable 16: National team reports on published and unpublished references on policy implementation and desertification at the local, regional and national level in project research countries*. Maastricht (NL): International Centre for Integrative Studies, pp. 32-45.

Oñate, J.J., Cummings, C., Gómez, A., Peco, B. and J.M. Sumpsi 2002a: Report on the results of in-depth qualitative interviews with local, regional and national policy-makers on the implementation and effects of past policies in the Guadalentín target area. In: Juntti, M. and G.A. Wilson (eds): *MEDACTION Project Deliverable 19: National team reports discussing the results of the in-depth qualitative interviews with local, regional and national policy-makers on the implementation and effects of past policies in each of the four target areas*. Maastricht (NL): International Centre for Integrative Studies, pp. 11-48.

Oñate J.J. and B. Peco 2005: Policy impact on desertification: stakeholders' perceptions in southeast Spain. *Land Use Policy* 22 (2): 103-114.

Oñate, J.J., Pereira, D., Suárez, F., Rodríguez, J.J. and J. Cachón 2002b: *Evaluación ambiental estratégica: la evaluación ambiental de políticas, planes y programas*. Madrid: Editorial Mundi-Prensa.

Oñate, J.J., Suárez, F. and J. Anula 2002c: Conservar el territorio y los paisajes: más allá de la Red Natura 2000. In: Araujo, J. (ed): *Ecología: perspectivas y políticas de futuro*. Sevilla (Spain): Junta de Andalucía-Fundación Alternativas, pp. 97-116.

Pahl, R.E. 1984: *Divisions of labour*. Oxford: Blackwell.

Pamo, E.T. 1998: Herders and wild game behaviour as a strategy against desertification in northern Cameroon. *Journal of Arid Environments* 39: 179-190.

Paniagua, A. 2001: European processes of environmentalization in agriculture: a view from Spain. In: Hoggart, K. and H. Buller (eds): *Agricultural transformation, food and environment: perspectives on European rural policy and planning (Vol 1)*. Aldershot: Ashgate, pp. 131-164.

Papanastasis, V.P. 1977: Fire ecology and management of *phrygana* communities in Greece. In: USDA/Forest Dervice (eds): *The environmental consequences of fire and fuel management in Mediterranean ecosystems*. Proceedings of the International Symposium held at Palo Alto, California, 1977. Washington (D.C.): USDA/Forest Service, pp. 476-482.

Papanastasis, V.P. 1998: Grazing intensity as an index of degradation in semi-natural ecosystems: the case of Psilorites Mountain in Crete. In: Enne, G., D' Angelo, M. and C. Zanolla (eds): *Indicators for assessing desertification in the Mediterranean*. Proceedings of the International Seminar held in Porto Torres, Italy, 18-20 September 1998. Rome: Osservatorio Nazionale sulla Desertificazione, pp. 146-158.

Papanastasis, V.P. and A. Giannakopoulos 1989: *A study of meadowland pasture development at Agrafa in Evrytania* (in Greek). Athens: Directorate of Studies and Programming [ATE].

Peco, B., Suárez, F., Oñate, J.J., Malo, J.E. and J. Aguirre 2000: Spain: first tentative steps towards an agri-environmental programme. In: Buller, H., Wilson. G.A. and A. Höll (eds): *Agri-environmental policy in the European Union*. Aldershot: Ashgate, pp. 145-168.

Peixoto, T.M. 1998: *Análise do impacto da Agenda 2000 na agricultura da Região do Alentejo*. Lisbon: Universidade Técnica de Lisboa [ISA].

References

Pérez, A. and F. Barrientos 1986: The Project 'LUCDEME' in southeast Spain to combat desertification in the Mediterranean region. In: Fantechi, R. and N.S. Margaris (eds): *Desertification in Europe*. Dordrecht (NL): Reidel.

Pérez-Sirvent, C., Martínez-Sánchez, M.J., Vidal, J. and A. Sánchez 2003: The role of low-quality irrigation water in the desertification of semi-arid zones in Murcia, SE Spain. *Geoderma* 113: 109-125.

Pérez-Trejo, F. 1992: *Desertification and land degradation in the European Mediterranean*. Luxembourg: European Commission (Report EUR 14850).

Pérez Yruela, M. 1995: Spanish rural society in transition. *Sociologia Ruralis* 35: 276-296.

Pescosolido, B.A. and B. Rubin 2000: The web of group affiliations revisited: social life, postmodernism and sociology. *American Sociological Review* 65: 52-76.

Phillips, C.P. 1998: The badlands of Italy: a vanishing landscape? *Applied Geography* 18 (3): 243-57.

Phillips, M. 1998: The restructuring of social imaginations in rural geography. *Journal of Rural Studies* 14 (2): 121-153.

Phillips, M. 2002: Distant bodies? Rural studies, political-economy and post-structuralism. *Sociologia Ruralis* 42 (2): 81-105

Pridham, G. 2002: National environmental policy-making in the European framework: Spain, Greece and Italy in comparison. In: Jordan, A. (ed): *Environmental policy in the European Union: actors, institutions and processes*. London: Earthscan, pp. 81-99.

Pridham, G. and D. Konstadakopoulos 1997: Sustainable development in Mediterranean Europe: interactions between European, national and sub-national levels. In Baker, S., Kousis, M., Richardson, P. and S. Young (eds): *The politics of sustainable development: theory, policy and practice within the EU*. London: Routledge, pp. 127-151.

Primdahl, J., Peco, B., Schramek, J. Andersen, E. and J.J. Oñate 2003: Environmental effects and effects measurement of agri-environmental policies. *Journal of Environmental Management* 67: 129-138.

Prior, L. 1997: Following in Foucault's footsteps: text and context in qualitative research. In: Silverman, D. (ed): *Qualitative research: theory, method and practice*. London: Sage, pp. 63-79.

Puigdefábregas, J. 1995: Desertification: stress beyond resilence - exploring a unifying process structure. *Ambio* 24: 311-313.

Puigdefábregas, J. and T. Mendizabal 1998: Perspectives on desertification: western Mediterranean. *Journal of Arid Environments* 39: 209-224.

Putnam, R. 1993: *Making democracy work: civic traditions in modern Italy*. Princeton: Princeton University Press.

RCDNM [Regional Conference on Desertification in the Northern Mediterranean] 1995: *Final report of the First Regional Conference on Desertification in Northern Mediterranean Countries*. Available online at http://www.unccd.int/regional/ northmed/meetings/regional/FinRep1995Almeria.pdf

Redclift, M., Shove, E., van der Meulen, B. and S. Raman 2000: *Social environmental research in the European Union: research networks and new agendas*. Cheltenham: Edward Elgar.

Rieutort, L. 1995: *L'élevage ovin en France: espaces fragiles et dynamique des systèmes agricoles*. Clermont-Ferrand: Presses Universitaires Blaise Pascal.

Robinson, G. 2004: *Geographies of agriculture: globalisation, restructuring and sustainability*. London: Pearson.

Rojo Serrano, L. 1998: Programmes of national agencies for mitigation of desertification in Spain. In: Burke, S. and J.B. Thornes (eds): *Actions taken by national governmental and non-governmental organisations to mitigate desertification in the Mediterranean*. Brussels: European Commission (Report EUR 18490 EN), pp. 211-232.

Romero-Díaz, A., Tobarra, P., López-Bermúdez, F. and G.G. Barberá 2002: Changing social and economical conditions in a region undergoing desertification in the Guadalentín. In: Geeson, N.A., Brandt, C.J. and J.B. Thornes (eds): *Mediterranean desertification: a mosaic of processes and responses*. Chichester: Wiley, pp. 289-302.

Rossi, L. and M. Iannetta 2002: Desertificazione: un fenomeno in espansione. *Agriculture* 1: 4-8.

Rossi-Doria, M. 2003: *Scritti sul Mezzogiorno*. Napoli: L'Ancora del Mediterraneo.

Roxo, M.J. and P.C. Casimiro 1998: Human impact on land degradation in the inner Alentejo, Mértola, Portugal. In: Mairota, P., Thornes, J.B. and N. Geeson (eds): *Atlas of Mediterranean environments in Europe: the desertification context*. Chichester: Wiley, pp. 106-109.

Roxo, M.J. and J.M. Mourão 1994: *Land degradation in the south interior Alentejo-Mértola region: historical overview of agricultural impacts on the environment*. Lisbon: Universidade Nova de Lisboa.

Rubio, J.L. 1995: Desertification: evolution of a concept. In Fantechi, R., D. Peter, P. Balabanis and J.L. Rubio (eds): *Desertification in a European context: physical and socio-economic aspects*. Proceedings of the European School of Climatology and Natural Hazards Course in Alicante, Spain, 6-13 October 1993. Luxembourg: Office for the Official Publications of the European Communities, pp. 5-13.

Sabatier, P.A. 1986: Top-down vs. bottom-up approaches to implementation research: a critical analyses and suggested synthesis. *Journal of Public Policy* 6 (1): 21-48.

Saraceno, E. 1994: Alternative readings of spatial differentiation: the rural versus the local approach in Italy. *European Review of Agricultural Economics* 21 (3/4): 451-474.

Saurí, D. and L. Del Moral 2001: Recent developments in Spanish water policy: alternatives and conflicts at the end of the hydraulic age. *Geoforum* 32: 351-362.

Seely, M.K. 1998: Can science and community action connect to combat desertification? *Journal of Arid Environments* 39: 267-277.

Senato della Repubblica 1998: *Indagine conoscitiva sulla difesa del suolo*. Rome: Senato della Repubblica.

Silverman, D. 1997: Qualitative research: theory, method and practice. London: Sage.

Stame, N. 2001: La vitalità dell'impresa familiare nel Mezzogiorno. In: Becattini, G., Bellandi, M., Dei Ottati, G. and F. Sforzi (eds): *Il caleidoscopio dello sviluppo locale: trasformazioni economiche nell'Italia contemporanea*. Torino: Rosenberg and Sellier, pp. 143-170.

Sullivan, S. 2000: Getting the science right, or introducing science in the first place? Local 'facts', global discourse: 'desertification' in northwest Namibia. In: Stott, P. and S. Sullivan (eds): *Political ecology: science, myths and power*. London: Arnold, pp. 15-44.

Sumpsi, J.M. 2001: Actors, institutions and attitudes to rural development: the Spanish national report. In: Baldock, D., Dwyer, J., Lowe, P. and N. Ward (eds): *The nature of rural development: towards a sustainable integrated rural policy in Europe*. London: IEEP.

Sumpsi, J.M., Garrido, A., Blanco, M., Varela, C. and E. Iglesias 1998: *Economía y política de gestión del agua en la agricultura*. Madrid: MAPA-Mundi Prensa.

Swift, J. 1996: Desertification: narratives, winners and losers. In: Leach, M. and R. Mearns (eds): *The lie of the land: challenging received wisdom on the African environment*. Oxford: Heinemann, pp. 73-90.

Thomas, D.S. 1997: Science and the desertification debate. *Journal of Arid Environments* 37: 599-608.

Thornes, J.B. 1993: *MEDALUS: executive summary*. Brussels: Commission of the European Communities.

Thornes, J.B. 1996: Introduction. In: Brandt, C.J. and J.B. Thornes (eds): *Mediterranean desertification and land use*. Chichester: Wiley, pp. 1-11.

Thornes, J.B. 1998a: Mediterranean desertification. In: Mairota, P., Thornes, J.B. and N. Geeson (eds): *Atlas of Mediterranean environments in Europe: the desertification context*. Chichester: Wiley, pp. 2-4.

Thornes, J.B. 1998b: Mediterranean desertification and DiCastri's 5th dimension. *Mediterraneo* 12/13: 149-166.

Thornes, J.B. and P. Mairota 1995: MEDALUS: achievements and prospects. In: Enne, G., Aru, A. and G. Pulina (eds): *Land use and soil degradation: MEDALUS in Sardinia*. Sassari: Università di Cagliari, pp. 1-13.

Trigilia, C. 1994: *Sviluppo senza autonomia*.Bologna: Il Mulino.

Troeh, F.R., Hobbs, J.A. and R.L. Donahue 2004: *Soil and water conservation for productivity and environmental protection (4th ed)*. New Jersey: Prentice Hall.

Tsoladikis, C. 1998: Report on research and action programmes for mitigation of desertification by non-governmental organisations in Greece. In: Burke, S. and J.B Thornes (eds): *Actions taken by national governmental and non-governmental organisations to combat desertification in the Mediterranean*. Brussels: European Commission (Report EUR 18490 EN), pp. 63-76.

UN [United Nations] 1992: *Agenda 21: programme of action for sustainable development*. New York: United Nations.

UN [United Nations] 1994: *The Convention to Combat Desertification*. New York and Paris: United Nations.

UNCCD [United Nations Convention on Desertification] 1994: *United Nations Convention to Combat Desertification*. Geneva: United Nations Environmental Programme.

UNCCD [United Nations Convention on Desertification] 2001: *Action plan to combat desertification in Africa*. Fact Sheet 11. http://www.unccd.int/publicinfo/factsheets/ showFS.php?number=11

UNCOD [United Nations Convention on Desertification] 1977: *Desertification: its causes and consequences*. New York: UNEP and Pergamon Press.

UNECE [United Nations Economic Commission for Europe] 1998: *Convention on access to information, public participation in decision-making and access to justice in environmental matters*. Aarhus: UNECE.

UNEP [United Nations Environmental Programme] 1991: *Status of desertification and implementation of the United Nations Plan of Action to Combat Desertification* Nairobi: UNEP (Report UNEP/GCSS.III/3).

UNEP [United Nations Environmental Programme] 1992: *World atlas of desertification*. London: Edward Arnold.

Valor, E. and V. Castellas 1995: The use of thermal infrared remote sensing in desertification studies: first results of the EFEDA, Hapex-Sahel and DEMON projects. In: Fantechi, R., Peter, D., Balabanis, P. and J.L. Rubio (eds): *Desertification in a European context: physical and socio-economic aspects*. Proceedings of the European School of Climatology and Natural Hazards Course in Alicante, Spain, 6-13 October 1993. Luxembourg: Office for the Official Publications of the European Communities, pp. 627-635.

Van der Leeuw, S.E. (ed) 1998: *The Archaeomedes Project: understanding the natural and anthropogenic causes of land degradation and desertification in the Mediterranean basin - research results*. Luxembourg: Office for Official Publications of the European Communities (Report EUR 18181).

Van der Leeuw, S.E. 1999: Degradation and desertification: some lessons from the long-term perspective. In: Balabanis, P., Peter, D., Ghazi, A. and M. Tsogas (eds): *Mediterranean desertification: research results and policy implications* (Vol 1). Luxembourg: Office for Official Publications of the European Communities, pp. 17-31.

Van der Ploeg, D. 1997: On rurality, development and rural sociology. In: De Haan, H. and N. Long (eds): *Images and realities of rural life*. Wageningen (NL): Wageningen Pers, pp. 39-73.

Van Rooyen, A.F. 1998: Combating desertification in the southern Kalahari: connecting science with community action in South Africa. *Journal of Arid Environments* 39: 285-297.

Van Wesemael, B., Cammeraat, E., Mulligan, M. and S. Burke 2003: The impact of soil properties and topography on drought vulnerability of rainfed cropping systems in southern Spain. *Agriculture, Ecosystems and Environment* 94: 1-15.

Vieira, L.M. and P. Eden 2000: Portugal: agri-environmental policy and the maintenance of biodiversity-rich extensive farming systems. In: Buller, H., Wilson, G.A. and A. Höll (eds): *Agri-environmental policy in the European Union*. Aldershot: Ashgate, pp. 203-218.

Vieira, L.M. and P. Eden 2002: Report on the results of in-depth qualitative interviews with local, regional and national policy-makers on the implementation and effects of past policies in the Alentejo target area. In: Juntti, M. and G.A. Wilson (eds): *MEDACTION Project Deliverable 19: National team reports discussing the results of the in-depth qualitative interviews with local, regional and national policy-makers on the implementation and effects of past policies in each of the four target areas*. Maastricht (NL): International Centre for Integrative Studies, pp. 72-91.

Villaret, A. 1996: *Eléments pour un bilan de la politique de la montagne*. Paris: La Documentation Française.

Ward, N., Lowe, P., Seymour, S. and J. Clark 1995: Rural restructuring and the regulation of farm pollution. *Environment and Planning A* 27: 1193-1211.

WCED [World Commission on Environment and Development] 1987: *Our common future*. Oxford: Oxford University Press.

Whitby, M. 1996 (ed): *The European environment and CAP reform*. Wallingford: CAB International.

Williams, M. 2001: Market reforms, technocrats and institutional innovation. *World Development* 30 (3): 395-412.

Wilson, G.A. 2001: From productivism to post-productivism and back again? Exploring the (un)changed natural and mental landscapes of European agriculture. *Transactions of the Institute of British Geographers* 26 (1): 77-102.

Wilson, G.A. 2002: 'Post-Produktivismus in der europäischen Landwirtschaft: Mythos oder Realität? *Geographica Helvetica* 57 (2): 109-126.

Wilson, G.A. 2004: The Australian *Landcare* movement: towards 'post-productivist' rural governance? *Journal of Rural Studies* 20: 461-484.

Wilson, G.A. and R.L. Bryant 1997: *Environmental management: new directions for the 21st century*. London: UCL Press.

Wilson, G.A. and H. Buller 2001: The use of socio-economic and environmental indicators in assessing the effectiveness of EU agri-environmental policy. *European Environment* 11: 297-313.

Wilson, G.A. and K. Hart 2000: Financial imperative or conservation concern? EU farmers' motivations for participation in voluntary agri-environmental schemes. *Environment and Planning A* 32 (12): 2161-2185.

Wilson, G.A. and K. Hart 2001: Farmer participation in agri-environmental schemes: towards conservation-oriented thinking? *Sociologia Ruralis* 41 (2): 254-274.

Wilson, G.A., Petersen, J.E. and A. Höll 1999: EU member state responses to Agri-Environmental Regulation 2078/92/EEC - towards a conceptual framework? *Geoforum* 30 (2): 185-202.

Wilson, G.A. and O.J. Wilson 2001: *German agriculture in transition: society, policies and environment in a changing Europe*. Basingstoke: Palgrave.

Winter, S. 1990: Integrating implementation research. In: Palumbo, D. and D. Calista (eds): *Implementation and the policy process: opening the black box*. Westport (Conn.): Greenwood Press, pp. 19-38.

Yassoglou, N. 1990: Desertification in Greece. In: Rubio, J.L. and R.J. Rickson (eds): *Strategies to combat desertification in Mediterranean Europe*. Brussels: Directorate General for Agriculture of the European Commission (Report EUR 11175), pp. 148-162.

Index